STATISTICS FOR ANALYTICAL CHEMISTRY
Third Edition

Ellis Horwood and Prentice Hall
are pleased to announce their collaboration in a new imprint whose list will encompass outstanding works by world-class chemists, aimed at professionals in research, industry and academia. It is intended that the list will become a by-word for quality, and the range of disciplines in chemical science to be covered are:

ANALYTICAL CHEMISTRY
ORGANIC CHEMISTRY
INORGANIC CHEMISTRY
PHYSICAL CHEMISTRY
POLYMER SCIENCE & TECHNOLOGY
ENVIRONMENTAL CHEMISTRY
CHEMICAL COMPUTING & INFORMATION SYSTEMS
BIOCHEMISTRY
BIOTECHNOLOGY

Ellis Horwood   PTR Prentice Hall
ANALYTICAL CHEMISTRY SERIES

Series Editors:
Dr Mary Masson
Dr Julian Tyson
Dr P Stockwell

Consultant Editors:
Professor Jim Miller
Dr Robert Chalmers
Ellis Horwood, M.B.E.

Current Ellis Horwood   PTR Prentice Hall
Analytical Chemistry Series titles

Guzman	**CAPILLARY ELECTROPHORESIS**
Miller & Miller	**STATISTICS FOR ANALYTICAL CHEMISTRY: 3rd Edition**
Smyth	**CHEMICAL ANALYSIS IN COMPLEX MATRICES**

STATISTICS FOR ANALYTICAL CHEMISTRY
Third Edition

J.C. MILLER
Wyggeston and Queen Elizabeth 1 College, Leicester

J.N. MILLER
Senior Pro-Vice Chancellor, Loughborough University
of Technology

ELLIS HORWOOD PTR PRENTICE HALL
NEW YORK   LONDON   TORONTO   SYDNEY   TOKYO   SINGAPORE

This Third Edition first published in 1993
Reprinted in 1993 and 1994 by
Ellis Horwood Limited
Market Cross House, Cooper Street
Chichester
West Sussex, PO19 1EB
A division of
Simon & Schuster International Group

Printed and bound in Great Britain by
Hartnolls, Bodmin

Library of Congress Cataloging-in-Publication Data

Available from the publisher

British Library Cataloguing in Publication Data

A catalogue record for this book is available from the British Library

ISBN   0-13-030990-7   (pbk)

1 2 3 4 5   97 96 95 94 93

Table of contents

Preface

To add yet another volume to the already numerous texts on statistics might seem to be an unwarranted exercise, yet the fact remains that many highly competent scientists are woefully ignorant of even the most elementary statistical methods. It is even more astonishing that analytical chemists, who practise one of the most quantitative of all sciences, are no more immune than others to this dangerous, but entirely curable, affliction. It is hoped, therefore, that this book will benefit analytical scientists who wish to design and conduct their experiments properly, and extract as much information from the results as they legitimately can. It is intended to be of value to the rapidly growing number of students specializing in analytical chemistry, and to those who use analytical methods routinely in everyday laboratory work.

There are two further and related reasons that have encouraged us to write this book. One is the enormous impact of microelectronics, in the form of microcomputers and hand-held calculators, on statistics: these devices have brought lengthy or difficult statistical procedures within the reach of all practising scientists. The second is the rapid development of new 'chemometric' procedures, including pattern recognition, optimization, numerical filter techniques, simulations and so on, all of them made practicable by improved computing facilities. The last chapter of this book attempts to give the reader at least a flavour of the potential of some of these newer statistical methods. We have not, however, included any computer programs in the book — partly because of the difficulties of presenting programs that would run on all the popular types of microcomputer, and partly because there is a substantial range of suitable and commercially available books and software.

The availability of this tremendous computing power naturally makes it all the more important that the scientist applies statistical methods rationally and correctly. To limit the length of the book, and to emphasize its practical bias, we have made no attempt to describe in detail the theoretical background of the statistical tests described. But we have tried to make it clear to the practising analyst which tests are appropriate to the types of problem likely to be encountered in the laboratory. There are worked examples in the text, and exercises for the reader at the end of each chapter. Many of these are based on the data provided by research papers published in *The Analyst*. We are deeply grateful to Mr. Phil Weston, the Editor, for allowing us thus to make use of his distinguished journal. We also thank our colleagues, friends and family for their forbearance during the preparation of the book; the sources of the statistical tables, individually acknowledged in the appendices; the Series Editor, Dr. Bob Chalmers; and our publishers for their efficient cooperation and advice.

J. C. Miller
J. N. Miller

April 1984

Preface to the second edition

The generally favourable reception accorded to the first edition of this book has encouraged us to prepare a revised and slightly expanded second edition. We are particularly grateful to those who pointed out errors in the first edition and indicated possible improvements. Professor D. Thorburn Burns has been most helpful in connection with the propagation of errors, and we also thank Mr. J. E. Colbeck for his constructive comments on nomenclature.

The principal changes introduced into this edition are a new chapter (4) which deals for the first time with collaborative trials and provides expanded coverage of sampling and of control charts. There is also additional material on type I and type II errors, the sequential use of significance tests, non-parametric methods, and non-linear calibration plots. Finally, the discussions of pattern recognition and factorial design have been extended, and more detailed solutions to some of the problems have been provided.

Again we thank our publishers; the Series Editors, Drs Bob Chalmers and Mary Masson; and Mr. Phil Weston, Editor of *The Analyst,* for their further help and cooperation in the production of this edition.

J. C. Miller
J. N. Miller

February 1987

Preface to the third edition

The third edition of this book shows several changes compared with its predecessors. Again we thank those who have written to us commenting on the second edition or pointing out errors and obscure passages in it. Amongst these correspondents, we are especially grateful to George Ekstrom of the National Laboratory for Agricultural Chemistry, Uppsala, Sweden.

We have retained the seven chapters of the second edition, but most of them contain significant changes and updates. In the last few years personal computers have gained in power and speed (and fallen in cost) to an astonishing degree, and many extra facilities are thus offered to any analytical chemist through a variety of good software packages. Many of our comments on calculation methods and practices (Chapters 1 and 2) have thus been updated. We retain our conviction that this greater ease of calculation increases rather than diminishes the need for a clear understanding of the principles of the underlying statistical methods, and this approach still motivates the whole volume. In some areas greater computer power has led to significant changes in statistical practice. Two such changes in particular are reflected in the greater length of this edition: the increased use of curvilinear regression methods (covered in Chapter 5), and the rapidly rising popularity of robust statistical methods (Chapter 6). Other topics in fundamental statistics that receive fuller treatment include outliers, both in univariate statistics (Chapter 3) and in calibration methods (Chapter 5); initial data analysis (IDA, Chapter 6); and experimental design and optimization (Chapter 7). To make room for these changes the coverage of some little used rapid and non-parametric methods has been removed from Chapter 6. We have naturally been tempted to expand greatly the treatment of some of the multivariate methods featured in Chapter 7. Multivariate calibration makes its appearance for the first time, but despite their much increased use (and misuse!) in computer-dominated laboratories we have decided again to limit our discussion of these 'chemometric' methods to an elementary and fairly qualitative level which does not involve the reader in a knowledge of matrix algebra. Finally, the literature of statistics, both in terms of general texts and in more specific applications to analytical chemistry, is developing rapidly, so the bibliographies at the end of each chapter in the book have been completely revised.

We are deeply grateful to our publishers, especially Ellis Horwood, Clive Horwood and Patricia Morrison, for their continued help and support; and to the Series Editors, Mary Masson and Julian Tyson, and Harpal Minhas, Editor of the Royal Society of Chemistry's journal *The Analyst*, for their cooperation in the production of this edition.

J. C. Miller
J. N. Miller
May 1992

Glossary of symbols

a	—	intercept of regression line
b	—	gradient of regression line
c	—	number of columns in two-way ANOVA
C	—	correction term in two-way ANOVA
F	—	the ratio of two variances
h	—	number of samples in one-way ANOVA
μ	—	arithmetic mean of a population
M	—	number of minus signs in Wald–Wolfowitz runs test
n	—	sample size
N	—	number of plus signs in Wald–Wolfowitz runs test
N	—	total number of measurements in two-way ANOVA
ν	—	number of degrees of freedom
$P(r)$	—	probability of r
Q	—	Dixon's Q, used to test for outliers
r	—	product–moment correlation coefficient
r	—	number of rows in two-way ANOVA
r	—	number of smallest and largest observations omitted in trimmed mean calculations
R^2	—	coefficient of determination
R'^2	—	adjusted coefficient of determination
ρ	—	Spearman rank correlation coefficient
s	—	standard deviation of a sample
$s_{y/x}$	—	standard deviation of y-residuals
s_b	—	standard deviation of slope of regression line
s_a	—	standard deviation of intercept of regression line
$s_{(y/x)\mathrm{w}}$	—	standard deviation of y-residuals of weighted regression line
s_{x_0}	—	standard deviation of x-value estimated using regression line
s_{B}	—	standard deviation of blank
$s_{x_{\mathrm{E}}}$	—	standard deviation of extrapolated x-value
$s_{x_{0\mathrm{w}}}$	—	standard deviation of x-value estimated by using weighted regression line

σ	—	standard deviation of a population
σ_0^2	—	measurement variance
σ_1^2	—	sampling variance
t	—	quantity used in the calculation of confidence limits and in significance testing of mean (see Section 2.4)
T	—	grand total in ANOVA
T_1 and T_2	—	test statistics used in the Wilcoxon rank sum test
w	—	range
w_i	—	weight given to point on regression line
$\bar{x}$	—	arithmetic mean of a sample
x_0	—	x-value estimated by using regression line
x_0	—	outlier value of x
$\tilde{x}_i$	—	pseudo-value in robust statistics
x_E	—	extrapolated x-value
$\bar{x}_W$	—	arithmetic mean of weighted x-values
χ^2	—	quantity used to test for goodness-of-fit
$\hat{y}$	—	y-values predicted by regression line
$\bar{y}_W$	—	arithmetic mean of weighted y-values
y_B	—	signal from blank
z	—	standard normal variable

1

Introduction

1.1 ANALYTICAL PROBLEMS

An analytical chemist may be presented with two types of problem. Sometimes he is asked to provide only a qualitative answer. For example, the presence of boron in distilled water is very damaging in the manufacture of microelectronic components — "Does this distilled water sample contain any boron?" Again, the comparison of soil samples is a common problem in forensic science — "Could these two soil samples have come from the same site?" In other cases the problems posed are quantitative ones. "How much albumin is there in this sample of blood serum?" "How much lead in this sample of tap-water?" "This steel sample contains small quantities of chromium, tungsten and manganese — how much of each?" These are typical examples of single-component or multiple-component quantitative analyses.

Modern analytical chemistry is overwhelmingly a **quantitative** science. It is apparent that in many cases a quantitative answer will be much more valuable than a qualitative one; it may be useful for an analyst to be able to say that he has detected some boron in a distilled water sample, but it is much more useful for him to be able to say *how much* boron is present. The person who requested the analysis could, armed with this quantitative answer, judge whether or not the level of boron was of concern, consider how it might be reduced, and so on: but if he knew only that *some* boron was present it would be hard for him to judge the significance of the result. In other cases, it is only a quantitative result that has any value at all. For example, virtually all samples of (human) blood serum contain albumin; the only question is, how much?

It is important to note that even where a qualitative answer is required, quantitative methods are used to obtain it. This point is illustrated with the aid of the examples given at the beginning of this section. In reality, an analyst would never simply say "I can/cannot detect boron in this water sample". He would use a quantitative method capable of detecting boron at, say, levels of 1 μg/ml. If the test gives a negative result, it would then be described in the form "This sample contains less than 1 μg/ml boron". If the test gives a positive result the sample will be reported to contain at least 1 μg/ml boron. Much more complex quantitative approaches might be used to compare two soil samples. For example, the samples might be subjected to a particle size analysis, in which the proportions of the soil particles falling within a number, say 10, of particle-size ranges are determined. Each sample would then be characterized by these 10 pieces of data. Quite complex procedures (see Chapter 7) can then be used to provide a quantitative assessment of their similarity, and an estimate of the probability of the samples having a common origin can be given if comparable data from soil samples *known* to match are available.

1.2 ERRORS IN QUANTITATIVE ANALYSIS

Once we accept that quantitative studies will play a dominant role in any analytical laboratory, we must accept also that the errors that occur in such studies are of supreme importance. Our guiding principle will be that *no quantitative results are of any value unless they are accompanied by some estimate of the errors inherent in them.* This principle naturally applies not only to analytical chemistry but to any field of study in which numerical experimental results are obtained. We can readily examine a number of simple examples which not only illustrate the principle but also foreshadow the types of statistical problem we shall meet and solve in subsequent chapters.

Suppose a chemist synthesizes an analytical reagent which he believes to be entirely new. He studies it by using a spectrometric method and the compound gives a value of 104 (normally, most of our results will be cited in carefully-chosen units, but in this hypothetical example purely arbitrary units can be used). On checking the reference books, the chemist finds that no compound previously discovered has yielded a value of more than 100 when studied by the same method under the same experimental conditions. The question thus naturally arises, has our chemist really discovered a new compound? The answer to this question lies of course in the degree of reliance that we can place on that experimental value of 104. What errors are associated with it? If further study indicates that the result is correct to within 2 (arbitrary) units, i.e. the true value probably lies in the range 104 ± 2, then a new material has probably been characterized. If, however, investigations show that the error may amount to 10 units (i.e. 104 ± 10), then it is quite likely that the true value is actually less than 100, in which case a new discovery is far from certain. In other words, a knowledge of the experimental errors is crucial (in this case as in every other) to the proper interpretation of the results. In statistical terms this example would involve the comparison of the experimental result with an assumed or reference value: this topic is studied in detail in Chapter 3.

A more common situation is that of the analyst who performs several replicate determinations in the course of a single analysis. (The value and significance of such replicates is discussed in detail in the next chapter). Suppose an analyst performs a titrimetric experiment four times and obtains values of 24.69, 24.73, 24.77 and 25.39 ml. The first point to notice is that titration values are reported to the nearest 0.01 ml; this point is also discussed further in Chapter 2. It is also immediately apparent that all four values are different, because of the errors inherent in the measurements, and that the fourth value (25.39 ml) is substantially different from the other three. The question that arises here is — can this fourth value be safely rejected, so that (for example) the mean titre is reported as 24.73 ml? In statistical terms, is the value 25.39

ml an 'outlier'? The important topic of outlier rejection is discussed in detail in Chapter 3.

Another frequent problem involves the comparison of two (or more) sets of results. Suppose that an analyst measures the vanadium content of a steel sample by two separate methods. With the first method he obtains an average value of 1.04%, with an estimated error of 0.07%: using the second method, he obtains an average value of 0.95% and an error of 0.04%. Several questions arise from the comparison of these results. Are the two average values significantly different, or are they indistinguishable within the limits of experimental error? Is one method significantly less error-prone than the other? Which of the mean values is actually closer to the truth? Again, Chapter 3 discusses these and related questions.

To conclude this section we note that many analyses are based on graphical methods. Instead of making repeated measurements on the same sample, we perform a series of measurements on a small group of standards which have known concentrations covering a considerable range. In this way we set up a calibration curve that can be used to estimate the concentrations of test samples studied by the same procedure. In practice, of course, all the measurements (those utilizing the standards and those on the test samples) will be subject to errors. It is necessary, for example, to assess the errors involved in drawing the calibration curve; to estimate the error in the concentration of a single sample determined by using the curve; and to estimate the limit of detection of the method i.e. the smallest quantity of analyte that can be detected with a particular degree of confidence. These procedures, which are especially commonplace in instrumental analysis, are described in Chapter 5.

These examples represent only a fraction of the possible problems arising from the occurrence of experimental errors in quantitative analysis. As we have seen, however, the problems have to be solved if the quantitative data are to have any real meaning. It is thus apparent that we must study the various types of error in more detail.

1.3 TYPES OF ERROR

Experimental scientists make a fundamental distinction between three types of error. These are known as **gross**, **random**, and **systematic** errors. Gross errors are readily described: they may be defined as errors which are so serious that there is no real alternative to abandoning the experiment and making a completely fresh start. Examples would include a complete instrument breakdown, accidentally dropping or discarding a crucial sample, or discovering during the course of the experiment that a supposedly pure reagent was in fact badly contaminated. Such errors (which occur occasionally even in the best-regulated laboratories!) are normally very easily recognized. In our discussion we thus have only to distinguish carefully between **random** and **systematic** errors.

We can best make this distinction by careful study of a real experimental situation. Four students (A–D) each perform an analysis in which *exactly* 10.00 ml of *exactly* 0.1M† sodium hydroxide is titrated with *exactly* 0.1M hydrochloric acid. Each student performs five replicate titrations, with the results shown in Table 1.1.

† The symbol 1M stands for 1 mole of the specified material per litre of solution.

The results obtained by student A have two important characteristics. First, they are all very close; all the results lie between 10.08 and 10.12 ml. In everyday terms we would say that the results are highly *reproducible*. The second feature of the results is that they are *all too high*: in this experiment (somewhat unusually) we know the correct answer in advance — it should of course be 10.00 ml. It is evident that two entirely separate types of error have occurred within this student's experiment. First, there are **random errors** — *these cause the individual results to fall on both sides of the average value* (10.10 ml in this case). Statisticians say that random errors affect the **precision**, or **reproducibility**, of an experiment. In the case of student A it is clear that the random errors are small, so we say that the results are **precise**. In addition, however, there are **systematic errors** — *these cause all the results to be in error in the same sense* (in this case they are all too high). Systematic errors, also called **bias**, affect **accuracy**, i.e. *proximity to the true value*. In many experiments the random and systematic errors are not only readily distinguishable by inspection of the results, they also have quite distinct origins in terms of experimental technique and equipment. Before we examine the causes of the errors in this experiment, however, we can discuss briefly the results obtained by students B–D. Student B has obtained results which are in direct contrast to those of student A. The average of the five results (10.01 ml) is very close to the true value, so we could characterize the data as accurate — without substantial systematic error. The spread of the results is very large, however, indicating poor precision and the presence of substantial random errors. Comparison of these results with those obtained by student A shows clearly that random and systematic errors can occur independently of one another. This conclusion is reinforced by the data of students C and D. Student C's work is neither precise (range 9.69–10.19 ml) nor accurate (average 9.90 ml). Student D has achieved both precise (range 9.97–10.04 ml) and accurate (average 10.01 ml) results. The distinction between random and systematic errors, and precision and accuracy respectively, is summarized in Fig. 1.1.

The apparently simple experiments described in the previous paragraphs merit several additional comments. It is especially noteworthy that the words **precision** and **accuracy** have entirely different meanings in the theory of errors, whereas they are used interchangeably in everyday English. Several dictionaries, for example, cite accuracy as a definition of precision (and vice versa!). This unfortunate confusion undoubtedly makes it more difficult to remember the essential difference between random and systematic errors, to which the words precision and accuracy respectively refer. Two further points of nomenclature may be mentioned here. Some texts speak of 'determinate' and 'indeterminate' errors, meaning the same as systematic and random errors respectively. Also, although we earlier used the word 'reproducibility' as an approximate definition of precision, modern convention makes a careful distinction between **reproducibility** and **repeatability**.

We can illustrate this distinction by an extension of our previous experiment. In the normal way student A (for example) would do the five replicate titrations in rapid succession; very probably the whole exercise would not take more than an hour or so. The same set of solutions and the same glassware would be used throughout, the same preparation of indicator would be added to each titration flask, and the temperature, humidity and other laboratory conditions would remain much the same. In such circumstances the precision measured would be the *within-run*

Table 1.1. Random and systematic errors

Student	Results (ml)	Comment
A	10.08 10.11 10.09 10.10 10.12	Precise Inaccurate
B	9.88 10.14 10.02 9.80 10.21	Accurate Imprecise
C	10.19 9.79 9.69 10.05 9.78	Inaccurate Imprecise
D	10.04 9.98 10.02 9.97 10.04	Accurate Precise

precision: this is called the **repeatability**. Suppose, however, that for some reason the same student did the titrations on five different occasions. In such circumstances he or she would probably use different pieces of glassware and different batches of indicator; the laboratory conditions might also change from time to time. So it would not be surprising to find a greater spread of the results in this case. This set of data would reflect the *between-run* precision of the method, and it is this to which the term **reproducibility** should be applied. Table 1.2 summarizes the definitions of the terms used thus far and the relationships between them.

One further lesson may be learned from the titration experiment. It is easy to appreciate that the data obtained by student C are unacceptable, and that those of student D are the most acceptable. Sometimes, however, it may happen that two methods are available for a particular analysis, one of which is believed to be precise but inaccurate, and the other accurate but imprecise. In other words we may have to choose between the types of results obtained by students A and B respectively. Which type of result is preferable? It is impossible to give a dogmatic answer to this question, if only because the choice of analytical method will often in practice be based on the cost, ease of automation, speed of analysis, and other factors falling outside our elementary evaluation. It is nonetheless important to realize that a method which is substantially free from systematic errors may still, if it is very imprecise, give an average value that is (by chance) considerably removed from the correct value. On the other hand a method that is precise but inaccurate (cf. student A) can be converted into one that is both precise *and* accurate (cf. student D) if the systematic errors can be discovered and hence removed. There will also be cases where, because the measurements being attempted are completely new, no check for systematic errors is feasible. Random errors can never be eliminated, though by

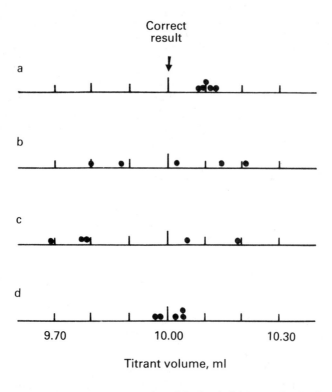

Fig. 1.1.   Accuracy and precision — dot plots of the data in Table 1.1.   In (a) the data are precise but inaccurate, in (b) accurate but imprecise, in (c) inaccurate and imprecise, and in (d) accurate and precise.

Table 1.2.   Summary of definitions

Types of error	
Random	Systematic
Affect precision	Affect accuracy
Within-run precision is **repeatability**	Proximity to the truth
Between-run precision is **reproducibility**	
Also known as indeterminate errors	Also known as determinate errors, bias

careful technique we can minimize them, and by making repeated measurements we can measure them and evaluate their significance. Systematic errors can in many cases be removed by careful checks on our experimental technique and equipment.

This additional crucial distinction between the two major types of error is further explored in the next section.

1.4   RANDOM AND SYSTEMATIC ERRORS IN TITRIMETRIC ANALYSIS

The example of the students' titrimetric experiment showed clearly that random and systematic errors can occur independently of one another and thus presumably arise at different stages of the experiment. Since titrimetry is a relatively simple and still widely adopted procedure it is of interest and value to examine it in detail in this context. A complete titrimetric analysis can be regarded as having the following steps (some of which were omitted from the example in the preceding section in the interest of simplicity).

(1) Making up a standard solution of one of the reactants. This involves (a) weighing a weighing bottle or similar vessel containing some solid material, (b) transferring the solid material to a standard flask and weighing the bottle again to obtain by subtraction the weight of solid transferred, and (c) filling the flask up to the mark with water (assuming that an aqueous titration is contemplated).

(2) Transferring an aliquot of the standard material to a titration flask with the aid of a pipette. This involves (a) filling the pipette to the appropriate mark, and (b) draining it in a specified manner into the titration flask.

(3) Titrating the liquid in the flask with a solution of the other reactant, added from a burette. This involves (a) filling the burette and allowing the liquid in it to drain until the meniscus is at a constant level, (b) adding a few drops of indicator solution to the titration flask, (c) reading the initial burette volume, (d) adding liquid to the titration flask from the burette a little at a time until the end-point is adjudged to have been reached, and (e) measuring the final level of liquid in the burette.

Even an elementary analysis of this type thus involves some ten separate steps, the last seven of which are normally, as we have seen, repeated several times. In principle, we should examine each step to evaluate the random and systematic errors which might occur. In practice, it is simpler to examine separately those stages which utilize weighings [steps 1(a) and 1(b)], and the remaining stages involving the use of volumetric equipment. The reader should notice that it is not the intention to give detailed descriptions of the experimental techniques used in the various stages. Similarly, methods for calibrating weights, glassware, etc. will not be given. (These topics are covered in adequate detail by the books listed in the bibliography at the end of the chapter). Important amongst the contributions to the errors are the tolerances of the weights used in the gravimetric steps, and of the volumetric glassware. Standard specifications for these tolerances are issued by such bodies as the British Standards Institution (BSI) and the American Society for Testing and Materials (ASTM). Some typical tolerance values are given in Table 1.3. The tolerance of a top-quality 100-g weight can be as low as ± 0.25 mg, although for a weight used in routine work the tolerance would be up to four times as large. Similarly the tolerance for a grade A 250-ml standard flask is ± 0.12 ml: grade B glassware generally has tolerances twice as large as grade A glassware. If a weight or a piece of glassware is within the tolerance limits, but not of exactly the correct weight or volume, a systematic error will arise. Thus, if the standard flask actually has a volume of 249.95 ml, this error will be reflected in the results of all the experiments

Table 1.3. Examples of tolerances of weights and glassware

Item	Tolerance
Weights (Class 1)	
100 g	±0.25 mg
1 g	0.034 mg
10 mg	0.010 mg
Weights (Class 3)	
100 g	±1.0 mg
1 g	0.10 mg
10 mg	0.030 mg
Glassware (Grade A)	
50 ml measuring cylinder	±0.25 ml
250 ml standard flask	0.12 ml
25 ml pipette	0.03 ml
50 ml burette	0.05 ml

(Data from ASTM Standard Specifications E617–78 and E694–79; see bibliography).

based on the use of that flask. Repetition of the experiment will not reveal the error; in each replicate the volume will be assumed to be 250.00 ml when in fact it is less than this. If, however, the results of an experiment using this flask are compared with the results of several other experiments (e.g. in other laboratories) done with other flasks, then the fact that all the flasks will have slightly different volumes will contribute to the random variation: it will be recalled that between-run random errors are said to affect the reproducibility of the results.

Weighing procedures are normally associated with very small **random** errors. In routine laboratory work a 'four-place' balance is commonly used, and the random error involved should not be greater than ±0.0001–0.0002 g (the next chapter describes in detail the statistical terms used to express random errors). Given that the quantity being weighed is normally of the order of 1 g or more, it is evident that the random error, expressed as a percentage of the weight involved, is not more than 0.02%. A good standard material for volumetric analysis should (amongst other characteristics) have as high a formula weight as possible, so as to minimize these random weighing errors when a solution of a given molarity is being made up. In some analyses 'microbalances' are used to weigh quantities of a few milligrams — but the errors involved are likely to be only a few micrograms.

Systematic errors in weighings can be appreciable, and have a number of well-established sources. These include adsorption of moisture on the surface of the weighing vessel; failure to allow heated vessels to cool to the same temperature as the balance before weighing (this error is especially commonplace in gravimetry when crucibles are weighed); corroded or dust-contaminated weights; and the buoyancy effect of the atmosphere, which acts to a different extent on objects of different density. For the most accurate work, weights must be calibrated against standards furnished by statutory bodies and standards authorities (see above). This calibration can be very accurate indeed, e.g. to ±0.01 mg for weights in the range 1–10 g. The buoyancy effect can be substantial. For example Skoog and West (see Bibliography) show that a sample of an organic liquid of density 0.92 g/ml which weighs 1.2100 g in air would weigh 1.2114 g in vacuo, a difference of more than 0.1%. Apart from the use of calibration procedures, which are further discussed in the next section, some simple experimental precautions can be taken to minimize these systematic errors. The most important of these is to perform all weighings **by difference**, i.e. a weighing bottle is weighed first with the sample contained in it, and secondly with the sample removed. The **difference** between the two weights gives the weight of the sample, and also cancels the systematic errors arising from (for example) the moisture and other contaminants on the surface of the bottle. (See also Section 2.8). *If such precautions are taken*, the errors in the weighing steps will be small, and it is probable that in most volumetric experiments weighing errors will be negligible compared with the errors arising from the use of volumetric equipment. Indeed, gravimetric methods are generally used for the calibration of volumetric glassware, by weighing (in standard conditions) the water that it contains or delivers; this procedure is obviously only valid because gravimetric errors are negligible compared with volumetric ones.

The random errors in volumetric procedures arise in the use of volumetric glassware. In filling a 250-ml standard flask to the mark, the error (i.e. the distance between the meniscus and the mark) might be about ±0.03 cm in a flask neck of diameter ca. 1.5 cm. This corresponds to a volume error of only about 0.05 ml — only 0.02% of the total volume of the flask. Similarly the random error in filling a 25-ml transfer pipette should not exceed 0.03 cm in a stem of diameter 0.5 cm; this gives a volume error of ca. 0.006 ml, 0.024% of the total volume. The error in reading a burette (of the conventional variety graduated in 0.1 ml divisions) is perhaps 0.01–0.02 ml. Each titration involves two such readings (the errors of which are *not* simply additive — see Chapter 2); if the titration volume is ca. 25 ml, the percentage error is again very small. It is clear that the experimental conditions should be arranged so that the volume of titrant is not too small (say not less than 10 ml), otherwise the errors will become appreciable. (This precaution is analogous to choosing a standard compound of high formula weight, to minimize the weighing error).Even though a volumetric analysis involves several steps, in each of which a piece of volumetric glassware is used, it is apparent that the random errors should be small if the experiments are performed with care. In practice a good volumetric analysis should have a relative standard deviation (see Chapter 2) of not more than about 0.1%. (Until recently, such precision was not normally attainable in instrumental analysis methods — and is still not commonplace — and this constituted one of the main advantages of 'classical' analysis techniques). In skilled hands, with all precautions taken, classical methods can give results with relative standard deviations as low as 0.01%.

Volumetric procedures incorporate several important sources of systematic error. Chief amongst these are the drainage errors in the use of volumetric glassware, calibration errors in the glassware, and 'indicator errors'. Perhaps the commonest error in routine volumetric analysis is to fail to allow enough time for a pipette to drain properly, or a meniscus level in a burette to stabilize. It is also worth remembering that pipettes are of two types, those which are emptied by drainage, and blow-out pipettes from which the last remaining liquid must be forcibly expelled. To confuse the two types, for example by blowing out a drainage pipette, would certainly count as a gross error! Drainage errors have a systematic as well as a random effect — the volume delivered is invariably less than it should be. The temperature at which an experiment is performed has two effects. Volumetric equipment is conventionally calibrated at 20°C, but the temperature in an analytical laboratory may easily be several degrees different from this, and many experiments, for example in biochemical analysis, are carried out in 'cold rooms' at ca. 4°C. The temperature affects both the volume of the glassware and the density of liquids. The coefficient of expansion for dilute aqueous solutions is about 0.025% per degree, whereas a soda-glass vessel will change in volume by about 0.003% per degree and a borosilicate glass vessel by about 0.001% per degree. It is evident that the changes in the volumes of glassware will only be important in work of the highest accuracy, and even then only if the temperature is very different from 20°C. Furthermore, the effects of the expansion of the solutions will be largely self-cancelling if all the solutions are maintained at the same temperature. The effect is much more marked with non-aqueous solutions.

Indicator errors can be quite substantial — perhaps larger than the random errors in a typical titrimetric analysis. For example, in the titration of $0.1M$ hydrochloric acid with $0.1M$ sodium hydroxide, we expect the end-point to correspond to a pH of 7. In practice, however, we estimate this end-point by the use of an indicator such as Methyl Orange. Separate experiments show that this substance changes colour over the pH range ca. 3–4. If, therefore, the titration is performed by adding alkali to acid, the indicator will yield an apparent end-point when the pH is ca. 3.5, i.e. just before the true end-point. The systematic error involved here is likely to be as much as 0.2%. If the titration is performed in reverse, i.e. by adding acid to alkali, the end-point indicated by the Methyl Orange will actually be a little beyond the true end-point, because of the need to achieve an acid pH to effect the colour change. Thus the systematic indicator error can be either positive or negative according to the design of the experiment; with a given procedure it will be either always positive or always negative. In either case the error can be evaluated and corrected by performing a **blank** experiment, i.e. by determining how much alkali (or acid) is required to produce the indicator colour change in the *absence* of the acid (alkali). A seldom mentioned error is the 'last drop' error. It is not known *how much* of the last drop added in a titration is actually needed to reach the end-point. Since either less or more than half the volume of the drop must be needed, this error can be reduced by subtracting half the volume of a drop.

In any analytical procedure, classical or instrumental, it is possible to consider and estimate the sources of random and systematic error arising at each separate stage of the experiment. It is very desirable for the analyst to do this, as it may enable him to avoid major sources of error by careful experimental design (see Sections 1.5 and 1.6). It is worth noting, however, that titrimetric analyses are rather unusual in that they involve no single step having an error which is far greater than the errors in the other steps. In many analyses the overall error is in practice dominated by the error in a single step: this point is further discussed in the next chapter.

1.5 HANDLING SYSTEMATIC ERRORS

Much of the remainder of this book will deal with the evaluation of random errors, which can be studied by a wide range of statistical methods. In most cases we shall assume for convenience that systematic errors are absent (though methods which test for the occurrence of systematic errors will be described). It is thus necessary at this stage to discuss systematic errors in more detail — how they arise, and how they may be countered. The example of the titrimetric analysis given in Section 1.3 shows clearly that systematic errors cause the mean value of a set of replicate measurements to deviate from the true value. It follows that (a) in contrast to random errors, systematic errors cannot be revealed merely by making repeated measurements, and that (b) unless the true result of the analysis is known in advance — a most unlikely situation! — very large systematic errors might occur, but go entirely undetected unless suitable precautions are taken. In other words, it is all too easy to completely overlook substantial sources of systematic error. A small number of examples will clarify both the possible problems and their solutions.

In recent years, much interest has been shown in the levels of transition metals in biological samples such as blood serum. Many determinations have been made of the levels of (for example) chromium in serum — with startling results. Different workers, all studying pooled serum samples from healthy subjects, have obtained chromium concentrations varying from <1 to ca. 200 ng/ml! In general the lower results have been obtained more recently, and it has gradually become apparent that the earlier, higher values were due at least in part to contamination of the samples by chromium from stainless-steel syringes, tube caps, and so on. The determination of traces of chromium, for example by atomic-absorption spectrometry, is in principle relatively straightforward, and no doubt each group of workers achieved results which seemed satisfactory in terms of precision, but in a number of cases the large systematic error introduced by the contamination was entirely overlooked. Methodological systematic errors of this kind are extremely common — incomplete washing of a precipitate in gravimetric analysis, and the indicator error in volumetric analysis (see Section 1.4) are further well-known examples.

Another class of systematic error that occurs widely arises when false assumptions are made about the accuracy of an analytical instrument. Experienced analysts know only too well that the monochromators in spectrometers gradually go out of adjustment, so that errors of several nanometers in wavelength settings are not uncommon, yet many photometric analyses are undertaken without appropriate checks being made. Very simple devices such as volumetric glassware, stop-watches, pH-meters and thermometers can all show substantial systematic errors, but many laboratory workers regularly use the instruments as though they are always perfectly accurate. Moreover, the increasing availability of instruments controlled by microprocessors or microcomputers has reduced to a minimum the number of operations and the level of skill required of their operators. In these circumstances the temptation to regard the instruments' results as beyond reproach is overwhelming — yet such instruments (unless they are 'intelligent' enough to be self-calibrating — see Section 1.7) are still subject to systematic errors.

Systematic errors arise not only from procedures or apparatus; they can also arise from human bias. Some chemists suffer from astigmatism or colour-blindness (the latter is more common amongst men than women) which might introduce errors into their readings of instruments and other observations. A number of authors have reported various types of number bias, for example a tendency to favour even over odd numbers, or 0 and 5 over other digits, in the reporting of results. It is thus apparent that systematic errors of several kinds are a constant, and often hidden, risk for the analyst, so the most careful steps to minimize them must be considered.

Several different approaches to this problem are available, and any or all of them should be considered in each analytical procedure.

The first precautions should be taken before any experimental work is begun. The analyst should consider carefully each stage of the experiment he is about to perform, the apparatus to be used and the sampling and analytical procedures to be adopted. He should try to identify at this earliest stage the likely sources of systematic error, such as the instrument functions that need calibrating, and the steps of the analytical procedure where errors are most likely to occur. He should also consider the checks he can subsequently make for systematic errors. Foresight of this kind can be immensely valuable (we shall see in the next section that similar advance attention should be given to the sources of random error) and is normally well worth the time invested. For example, a little thinking of this kind might well have revealed the possibility of contamination in the chromium determination described above.

The second line of defence against systematic errors lies in the design of the experiment at every stage. We have already seen (Section 1.4) that weighing by difference can remove some systematic gravimetric errors: these can be assumed to occur to the same extent in both weighings, so the subtraction process eliminates them. A further example of thoughtful experimental planning is provided by the spectrometer wavelength error described above. If the concentration of a sample of a single material is to be determined by absorption spectrometry, two procedures are possible. In the first, the sample is studied in a 1-cm path-length spectrometer cell at a single wavelength, say 400 nm, and the concentration of the test component is determined from the well-known equation $A = \varepsilon c l$ [where A, ε, c and l are respectively the measured absorbance, the accepted reference value of the molar absorptivity (units $l.mole^{-1}.cm^{-1}$) of the test component, the molar concentration of this analyte, and the path-length (cm) of the spectrometer cell]. Several systematic errors can arise here: the wavelength might, as already discussed, be (say) 405 nm rather than 400 nm, thus rendering the reference value of ε inappropriate; this reference value might in any case be wrong; the absorbance scale of the spectrometer might exhibit a systematic error; and the path-length of the cell might not be exactly 1 cm. Alternatively, the analyst might take a series of solutions of the test substance of *known* concentration, and measure the absorbance of each at 400 nm. (One of these calibrating solutions would be a blank, i.e. it would contain all the analytical reagents but none of the test substance — cf. Section 1.4). The results would then be used to construct a calibration graph for use in analysis of the test sample under exactly the same experimental conditions. This important approach to instrumental analysis is described in detail in Chapter 5. When this second method is used, the value of ε is not required, and the errors due to wavelength shifts, absorbance errors and path-length inaccuracies should cancel out, as they occur equally in the calibration and test experiments. Provided only that the conditions are indeed equivalent for the test and calibration samples (for example, we are assuming that the same cell is used and the wavelength and absorbance scales do not alter *during the experiment*) all the major sources of systematic error are in principle eliminated.

The final and perhaps most formidable protection against systematic errors is the use of standard reference materials and methods. Before the experiment is started, each piece of apparatus is calibrated by an appropriate procedure. We have already seen that volumetric equipment can be calibrated by the use of gravimetric methods.

Similarly, spectrometer wavelength scales can be calibrated with the aid of standard light-sources which have narrow emission lines at well-established wavelengths, and spectrometer absorbance scales can be calibrated with standard solid or liquid filters. In analogous fashion, most pieces of equipment can be calibrated so that their systematic errors are known in advance. The importance of this area of chemistry (and other experimental sciences) is reflected in the extensive work of bodies such as the National Physical Laboratory (in the U.K.), the National Institute for Science and Technology (NIST) (in the U.S.A.) and similar organizations elsewhere. Whole volumes have been written on the standardization of particular types of equipment (see bibliography), and a number of commercial organizations specialize in the scale of standard reference materials.

If systematic errors are occurring during chemical processes or as a result of using impure reagents, rather than in the equipment, an alternative form of comparison must be used, i.e. the determination must be repeated by use of an entirely independent procedure. If two (or more) chemically and physically unrelated methods are used to perform one analysis, and if they consistently yield results showing only random differences, it is a reasonable presumption that no significant systematic errors are present. For this approach to be valid, *each step* of the two experiments has to be independent. Thus in the case of serum chromium determinations, it would not be sufficient to replace the atomic-absorption spectrometry step by, for example, a colorimetric method or by plasma spectrometry: the systematic errors would only be revealed by altering the sampling methods also, e.g. by minimizing or eliminating the use of stainless-steel equipment. A further important point is that comparisons must be made over the whole of the concentration range for which an analytical procedure is to be used. For example, the widely-used Bromocresol Green dye-binding method for the determination of albumin in serum correlates well with alternative methods (e.g. immunological methods) at normal or high levels of albumin, but when the albumin levels are abnormally low (these are inevitably the cases of most clinical interest!) the agreement between the two methods is poor, the dye-binding method giving consistently (and erroneously) higher albumin concentrations. The statistical approaches used in method comparisons are described in detail in Chapters 3 and 5.

The prevalence of systematic errors in everyday analytical work is well illustrated by the results of **collaborative trials**. It is natural for an able and experienced analyst to believe that if, for example, he finds 10 ng/ml of a drug in a urine sample, then other analysts would obtain closely similar results for the same sample, any differences being due to random errors only. Unfortunately, this is far from true in practice. In many collaborative trials involving different laboratories, even when meticulously prepared and distributed aliquots of a single sample are examined by the same experimental procedures and the same types of instrument, the variation in the results often very greatly exceeds that which could reasonably be expected from random errors. The inescapable conclusion is that in many laboratories a variety of substantial systematic errors, both positive and negative, is going undetected or uncorrected. The obvious importance of this rather worrying situation, which has serious implications for all analytical scientists, has encouraged many studies of the methodology of collaborative trials, and of the statistical evaluation of their results. This topic is discussed in more detail in Chapter 4.

Though we have distinguished carefully thus far between random and systematic errors, it will be apparent that, in everyday analytical measurements, this distinction may become blurred. Whenever a procedure or instrument is checked for the presence of systematic errors, the checks themselves will be subjected to random errors, so the systematic error may not be identified or corrected perfectly! Thus the result of any analysis will be associated with both a random error and a systematic error which is not known exactly. This combination of errors has become known in recent literature as the *uncertainty* of the analytical result. It is clearly an awkward concept to handle: although random errors generally have a known distribution, and combine in a predictable way in a multi-stage experiment (see Chapter 2), the same is not true of systematic errors, so to give a quantitative estimate for the overall uncertainty in a result is far from simple. Despite these problems, the importance of the concept of uncertainty is clear, and it is likely that the next few years will see extensive efforts to extend and popularize this approach.

1.6 PLANNING AND DESIGN OF EXPERIMENTS

Many chemists regard statistical tests as methods to be used only to assess the results of completed experiments. While this is indeed a crucial area of application of statistics — we have seen that no quantitative result is worth anything if it is not accompanied by an estimate of its errors — it is also necessary to be aware of the importance of statistical concepts in the planning and design of experiments. In the previous section the value of trying to predict systematic errors in advance, thereby permitting the analyst to lay plans for countering them, was emphasized. The same considerations apply to random errors. As will be seen in Chapter 2, the combination of the random errors of the individual parts of an experiment to give an overall error requires the use of simple statistical formulae. In practice, the overall error is often dominated by the error in just one stage of the experiment, other errors having negligible effects when all the errors are combined correctly. Again it is obviously desirable to try to identify, *before the experiment begins*, where this single dominant error is likely to arise, and then to try to minimize it. Although random errors can never be eliminated, they can certainly be minimized by particular attention to experimental techniques: improving the precision of a spectrometric experiment by using a constant temperature sample cell would be a simple instance of such a precaution. For both random and systematic errors, therefore, the moral is clear — every effort must be made to identify the serious sources of error before practical work starts, so that the experiment can be designed to minimize such errors.

We must also give some preliminary consideration to a further, and more subtle, aspect of experimental design. In many analyses, one or more of the desirable features of the method (for example sensitivity, selectivity, sampling rate, low cost etc.) will be found to depend on a number of experimental factors. We shall wish to design the experiment so that the optimum combination of these factors is used, thereby obtaining the best sensitivity, selectivity etc. Although *some* preliminary experiments and prior knowledge will be involved, optimization should be performed (in the interests of conserving resources and time) before the method is put into routine or widespread use.

The complexity of optimization procedures can be illustrated with the aid of an example. In enzymatic analyses, the concentration of the analyte is determined from observations of the rate of an enzyme-catalysed reaction. The analyte is often the substrate, i.e. the compound that is changed in the reaction catalysed by the enzyme. Let us assume that we are seeking the maximum reaction rate in a particular experiment, and that the rate in practice depends upon (amongst other factors) the pH of the reaction mixture, the temperature, and the concentration of the enzyme. How are the optimum conditions to be found? It is easy to identify one possible approach. The analyst could perform a series of experiments, in each of which the enzyme concentration and the temperature are kept constant but the pH is varied. In each case the rate of the enzyme-catalysed reaction would be determined and an optimum pH value would thus be obtained — let us say that it is 7.5. A second series of reaction-rate experiments could then be performed, with the pH maintained at 7.5, the enzyme concentration again fixed, but the temperature varied. An optimum temperature would thus be found — say 40°C. Finally a series of experiments at pH 7.5 and 40°C but with various enzyme concentrations would indicate the optimum enzyme level. This approach to the optimization of the experiment is clearly tedious: in more realistic examples many more than three experimental factors might need investigation. It has the further and more fundamental objection that it assumes that the factors (pH, temperature, enzyme concentration) affect the reaction rate in an *independent* fashion. This might not be true. For example we have found that, at pH 7.5, the optimum temperature is 40°C. At a different pH, however, the optimum temperature might not be 40°C. In other words these factors may affect the reaction rate in an *interactive* fashion, and the conditions established in the series of experiments just described might not actually be the optimum ones. Had the first series of experiments been started in different conditions, a different set of 'optimum' values might have been obtained. It is apparent from this simple example that experimental optimization can be a formidable problem. This very important aspect of statistics as applied to analytical chemistry is considered in more detail in Chapter 7.

1.7 CALCULATORS AND COMPUTERS IN STATISTICAL CALCULATIONS

No chemist can be unaware of the astonishing developments that have recently taken place in the realm of microelectronics. These advances have made possible the construction of devices which enormously simplify statistical calculations. The rapid growth of 'chemometrics' — the application of mathematical methods to the solution of chemical problems of all types — is due to the ease with which large quantities of data can be handled, and complex calculations done, with calculators and computers.

These devices are available to the analytical chemist at several levels of complexity and cost, as follows.

(1) Hand-held calculators are extremely cheap, very reliable, and capable of performing many routine statistical calculations with a minimal number of keystrokes. In many calculators, preprogrammed functions will allow calculations of mean and standard deviation (see Chapter 2) and correlation and linear regression (see Chapter 5). Other calculators can be fitted with plug-in cards, or

programmed by the user, to perform additional calculations such as confidence limits (see Chapter 2), significance tests (see Chapter 3) and non-linear regression (see Chapter 5). For many applications in laboratories performing analytical research or routine analyses, calculators of these types will be more than adequate. Their disadvantages lie in their inability to handle very large quantities of data, and the problems from the limited number of significant figures incorporated into their algorithms (see Chapter 2).

(2) Microcomputers or personal computers (PCs), whose prices in real terms are falling almost as fast as their memory capacity and processor speeds are rising, are now found in all chemical laboratories, and most analytical scientists have their own PCs on their desks. Portable PCs facilitate the recording and calculation of data in the field, and are readily linked to their larger cousins on returning to the laboratory. All these devices have to be programmed to perform even the simplest statistical calculations, but a wealth of excellent software is available for all types of PC. The memory capacity and speed of PCs are now more than adequate for work with all but the largest data sets, though as with calculators problems may still arise with significant figures during statistical computations. PCs are routinely provided with word processors, which greatly aid the compilation of analytical reports and papers, and spreadsheet programs. The latter, though originally designed for financial calculations, are often more than adequate for simple statistical work, having several statistical functions (e.g. standard deviation or regression calculations) built in. The popularity of spreadsheets derives from their simplicity in use, and their ability to perform almost instant 'what if' calculations — for example, what would the mean and standard deviation of a set of results be *if* one suspect piece of data was omitted? Spreadsheets are designed to facilitate rapid data entry, and the spreadsheet data format is now generally used in more specialist suites of statistics software.

Perhaps the greatest value of PCs is that they can readily be interfaced to virtually all types of analytical instrument, with the dual roles of controlling the instrument's functions, and collecting and interpreting the resulting data: in many cases the data interpretation includes the production of results in graphical form, and even the printing of formal reports. Additional functions of the PC can include checking instrument performance, diagnosing and reporting malfunctions, storing large databases (e.g. of digitized spectra) and comparing analytical data with the databases, optimizing operating conditions (see Chapter 7), and selecting and using a variety of calibration calculations. As computer power increases, such developments are certain to continue: expert systems and neural networks are examples of artificial intelligence software already available in PC format. An increasing (though not always welcome) trend in instrumental analysis is for the measuring instrument to be supplied devoid of manual controls, because it is entirely controlled by its linked PC. A further concern is that the software provided with computer-controlled instruments is not always explained to the user: an analyst might thus suffer the misfortune of having a data set interpreted via (for example) a calibration routine which was not defined and might not always be appropriate. This is an undesirable situation, though the desire of instrument companies to try to protect costly software from piracy is understandable.

Yet another advantageous feature of PCs is that they can readily be 'networked', i.e. a group of PCs in the same or neighbouring laboratories can be linked so that both operating software and data can be freely passed from one to another. An obvious benefit of such networks is in the establishment of Laboratory Information Management Systems (LIMS) which allow large numbers of analytical specimens to be identified and tracked as they move through a laboratory. PCs attached to a range of instruments (often preceded by devices which use bar-coded labelling) can pass a range of different analytical results to a central computer which (for example) prints a summary report, including a statistical evaluation.

(3) Larger computers, the so-called mini-computers and main-frame systems, are now used in chemistry only for very large-scale calculations (e.g. X-ray crystallographic computations, advanced molecular modelling etc.), or for extensive data storage. Their use in statistical work is similarly confined to the study of exceptionally big data sets, or where numerous iterations are needed.

It is most important for the analytical chemist to remember that the availability of all these data-handling facilities increases rather than decreases the need for a sound knowledge of the principles underlying statistical calculations. A computer or calculator will rapidly perform any statistical test or calculation selected by the user, *whether or not that procedure is suitable for the data under study*. For example, a linear least-squares program will determine a straight line to fit *any* set of x and y values, even in cases where visual inspection would show at once that such a program is entirely inappropriate (see Chapter 5). Again, a simple program for testing the significance of the difference between the means of two data sets may assume that the variances (see Chapter 2) of the two sets are similar: but the program will blindly perform the calculation on request and provide a 'result' even if the variances actually differ significantly. Even quite comprehensive suites of computer programs often fail to provide advice on the choice of statistical method appropriate to a given set of data. The analyst must thus use his knowledge of statistics, and his common sense, to ensure that the correct calculation is performed.

In this book no attempt is made to provide listings of computer programs for statistical calculations. Excellent programs are available for many of the methods to be described in subsequent chapters, including the more advanced techniques such as non-linear regression (Chapter 5), pattern recognition, and simplex optimization (Chapter 7).

BIBLIOGRAPHY

In addition to the present text, several books are available which deal with applications of statistics in analytical chemistry. These include *Practical Statistics for Analytical Chemists* by Robert L. Anderson (Van Nostrand Reinhold, New York, 1987), and *Statistics for Analytical Chemists* by R. Caulcutt and R. Boddy (Chapman and Hall, London, 1983). An earlier but interesting text is *Statistical Theory and Methodology of Trace Analysis* by C. Liteanu and I. Rica (Ellis Horwood, Chichester, 1980).

Many conventional textbooks of analytical chemistry include sections or chapters on the statistical treatment of experimental results. These sections are very variable in their coverage, but a very good example is Chapter 3 of *Fundamentals of Analytical Chemistry* by D. A. Skoog and D. M. West, 4th Ed., Holt-Saunders, New York, 1982.

Many general statistics books have been written. Among them, *Facts from Figures* by M. J. Moroney (Penguin Books) is a long-standing and engagingly written classic. More recently published, and again very readable, is *Statistics in Action* by Peter Sprent (Penguin Books). A further large number of statistics texts are oriented towards scientific (not necessarily chemical) users; *Statistics for Technology* by Christopher Chatfield (Chapman & Hall, London) is an excellent example. Such books provide a quite comprehensive theoretical background to many of the methods described in the present volume.

All users of statistics require a good set of statistical tables. *Elementary Statistics Tables* by Henry R. Neave (Allen & Unwin, London) is strongly recommended, in view of its clear explanations of the tables and the associated statistical tests.

Several statutory and professional bodies in the U.K., the U.S.A. and elsewhere publish manuals and monographs on standard and recommended statistical procedures and on standard materials. Examples include the British Standards Institute (e.g. BS 5497, *Precision of Test Methods, Part 1 — Guide to the Determination of Repeatability and Reproducibility of a Standard Test Method;* and BS 1797, *Calibration of Volumetric Glassware):* the Association of Official Analytical Chemists (W. J. Youden and E. H. Steiner, *Statistical Manual of the AOAC*, 1975): the International Organisation for Standardisation (ISO) (e.g. IS 4259, *Petroleum Products — Determination and Application of Precision Data in Relation to Methods of Test):* and the American Society for Testing and Materials (ASTM) (e.g. Standard Specifications E617–78, *Laboratory Weights and Precision Mass Standards,* and E694–79, *Volumetric Ware.*

Monographs describing standard procedures and materials in specific areas of study are exemplified by A. Knowles and C. Burgess (eds.) *Standards in Absorption Spectrometry,* and J. N. Miller (ed.) *Standards in Fluorescence Spectrometry,* both published by Chapman & Hall in conjunction with the UV Spectrometry Group.

EXERCISES

1. A standard sample of pooled human blood serum contains 42.0 g of albumin per litre. Five laboratories (A–E) each do six determinations (on the same day) of the albumin concentration, with the following results (g/l. throughout):

A	42.5	41.6	42.1	41.9	41.1	42.2
B	39.8	43.6	42.1	40.1	43.9	41.9
C	43.5	42.8	43.8	43.1	42.7	43.3
D	35.0	43.0	37.1	40.5	36.8	42.2
E	42.2	41.6	42.0	41.8	42.6	39.0

Comment on the accuracy and precision of each of these sets of results.

2. Using the same sample and method as in question 1, laboratory A makes six further determinations of the albumin concentration, this time on six successive days. The values obtained are 41.5, 40.8, 43.3, 41.9, 42.2 and 41.7 g/l. Comment on these results.

3. The number of binding sites per molecule in a sample of monoclonal antibodies is determined four times, with results of 1.95, 1.95, 1.92 and 1.97. Comment on the accuracy and precision of these results.

4. Discuss the degrees of accuracy and precision desirable in the following analysis:

    (i) Determination of the lactate concentration of human blood samples.
    (ii) Determination of uranium in an ore sample.
    (iii) Determination of a drug in blood plasma after an overdose.
    (vi) Study of the stability of a colorimetric reagent by determination of its absorbance at a single wavelength over a period of several weeks.

5. For each of the following experiments, try to identify the major probable sources of random and systematic errors, and consider how such errors may be minimized:

    (i) The iron content of a large lump of ore is determined by taking a single small sample, dissolving it in acid, and titrating with ceric sulphate after reduction of Fe(III) to Fe(II).
    (ii) The same sampling and dissolution procedure is used as in (i) but the iron is determined colorimetrically after addition of a chelating reagent and extraction of the resulting coloured complex into an organic solvent.
    (iii) The sulphate content of an aqueous solution is determined gravimetrically with barium chloride as the precipitant.

2

Errors in classical analysis — statistics of repeated measurements

2.1 MEAN AND STANDARD DEVIATION

In this chapter some fundamental statistical concepts are introduced and applied to a situation common in classical analysis, i.e. repeated measurements of the same quantity are made. In Chapter 1 the various types of error were illustrated by considering the results of five replicate titrations done by each of four students: these results are reproduced below.

Student	Results (ml)				
A	10.08	10.11	10.09	10.10	10.12
B	9.88	10.14	10.02	9.80	10.21
C	10.19	9.79	9.69	10.05	9.78
D	10.04	9.98	10.02	9.97	10.04

Two criteria were used to compare these results, the average value and the degree of spread. The average value used was the **arithmetic mean**, $\bar{x}$, (usually abbreviated to the **mean**) which is the sum of all the measurements divided by the number of measurements:

$$\bar{x} = \sum_i x_i/n \qquad (2.1)$$

The most useful measure of spread is the **standard deviation**, s. This is defined by the formula:

$$s = \sqrt{\sum_i (x_i - \bar{x})^2/(n-1)} \qquad (2.2)$$

The calculation of the standard deviation of A's results is shown in Table 2.1. The reader should check that the standard deviations of the results of students B, C and D are 0.17, 0.21, and 0.033 ml respectively, giving quantitative confirmation of the assessments of precision made in Chapter 1.

Many pocket calculators will give the results of these calculations if the values of x are keyed in. However, care must be taken that the correct key is pressed to obtain the standard deviation. Some calculators give two different values for the standard deviation, one calculated by using Eq. (2.2) and the other calculated with the denominator of this equation, i.e. $(n-1)$, replaced by n. (The reason for these two different forms is explained below, p. 36.) Obviously, for large values of n the difference is negligible.

Unfortunately, in calculating a standard deviation, the calculator may round off numbers so that an erroneous value (even zero) may be obtained for the standard deviation. This usually happens when the difference between the input values occurs at the fourth or subsequent significant digit, depending on the make of calculator. For example, many calculators give the standard deviation of the three values 100.000, 100.001 and 100.002 as zero , whereas it is in fact 0.001. This problem can be overcome by **coding** the values, i.e. subtracting a fixed amount from each, say in this case 100.000, to give 0.000, 0.001 and 0.002. Since the standard deviation measures the spread about the mean, the standard deviation of these coded values is the same as that of the original values. (The mean of the original values is found by adding 100.000 to the mean of the coded values.)

The calculation in Table 2.1 was very simple because the values of $(x_i - \bar{x})^2$ could

Table 2.1.   Calculation of mean and standard deviation of A's results

	x_i(ml)	$(x_i - \bar{x})^2$
	10.08	0.0004
	10.11	0.0001
	10.09	0.0001
	10.10	0.0000
	10.12	0.0004
Total	50.50	0.0010

$$\bar{x} = 50.50/5 = 10.10 \text{ ml}$$
$$s = \sqrt{0.0010/4} = 0.016 \text{ ml}$$

be worked out mentally. This is not usually so and an alternative form of Eq. (2.2) can be used to simplify the arithmetic if a preprogrammed calculator is not available:

$$s = \sqrt{\frac{\sum_i x_i^2}{(n-1)} - \frac{(\sum_i x_i)^2}{n(n-1)}} \tag{2.3}$$

The mean and the standard deviation can also be calculated on a computer with a program of only a few lines. Such a program would only be of value if the computer itself was collecting the data or a large amount of data was involved. Example programs are given in the books in the bibliography at the end of Chapter 1.

The square of s is a very important statistical quantity known as the **variance**; its value will become apparent later in this chapter when we consider the propagation of errors. Also widely used is the **coefficient of variation (CV)**, also known as the **relative standard deviation (RSD)** and given by $100\,s/\bar{x}$. The CV or RSD, the units of which are obviously per cent, is an example of a **relative error**, i.e. an error estimate divided by an estimate of the absolute value of the measured quantity. Relative errors are frequently used in the comparison of the precisions of results which have different units or magnitudes, and are again important in calculations of error propagation.

2.2 DISTRIBUTION OF ERRORS

Although the standard deviation gives a measure of the spread of a set of results about the mean value, it does not indicate the way in which the results are distributed. To illustrate this we need a large number of measurements such as those in Table 2.2. This gives the results of 50 replicate determinations of the nitrate ion

Table 2.2.   Results of 50 determinations of nitrate ion concentration, in μg/ml

0.51	0.51	0.51	0.50	0.51	0.49	0.52	0.53	0.50	0.47
0.51	0.52	0.53	0.48	0.49	0.50	0.52	0.49	0.49	0.50
0.49	0.48	0.46	0.49	0.49	0.48	0.49	0.49	0.51	0.47
0.51	0.51	0.51	0.48	0.50	0.47	0.50	0.51	0.49	0.48
0.51	0.50	0.50	0.53	0.52	0.52	0.50	0.50	0.51	0.51

concentration in a particular water specimen, given to two significant figures. These can be summarized in a **frequency table** (Table 2.3). This table thus indicates that, in

Table 2.3.   Frequency table for measurements of nitrate ion concentration

Nitrate ion concentration (μg/ml)	Frequency
0.46	1
0.47	3
0.48	5
0.49	10
0.50	10
0.51	13
0.52	5
0.53	3

Table 2.2, the value 0.46 μg/ml appears once, the value 0.47 μg/ml appears three times, and so on. The reader can check that the mean of these results is 0.500 μg/ml and the standard deviation is 0.0165 μg/ml. (These values are arbitrarily given to three significant figures: further discussion of this important aspect of the presentation of results appears later in this chapter.) The distribution of the results can most easily be appreciated by drawing a **histogram** as in Fig.2.1. This shows that the

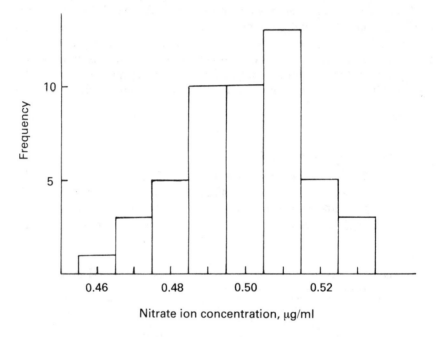

Fig. 2.1.   Histogram of the nitrate ion concentration data in Table 2.3.

distribution of the measurements is roughly symmetrical about the mean, with the measurements clustered towards the centre.

This set of 50 measurements is a **sample** from the very large (in theory infinite) number of measurements which we could make of the nitrate ion concentration. The set of possible measurements is called the **population**. *If there are no systematic errors*, then the mean of this population, denoted by μ, is the true value of the nitrate ion concentration which we are trying to determine. The mean, $\bar{x}$, of the sample gives us an estimate of μ. Similarly, the population has a standard deviation, denoted by σ. The value of the standard deviation, s, of the sample gives us an estimate of σ. Use of Eq. (2.2) gives us an unbiased estimate of σ. If n, rather than $(n-1)$, is used in the denominator of the equation the value of s obtained tends to underestimate σ (see p. 34 above).

The measurements of nitrate ion concentration given in Table 2.2 have only certain discrete values, because of the limitations of the method of measurement. In theory a concentration could take any value, so to describe the form of the population from which the sample was taken a continuous curve is needed. The mathematical model usually used is the **normal** or **Gaussian distribution** which is described by the equation

$$y = \frac{\exp\{-(x-\mu)^2/2\sigma^2\}}{\sigma\sqrt{2\pi}}$$
(2.4)

and its shape is shown in Fig. 2.2. There is no need to remember this complicated

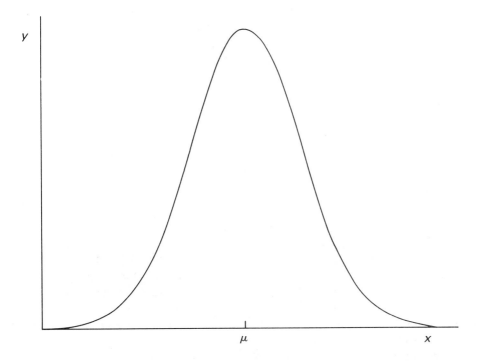

Fig. 2.2.   The normal distribution, $y = \exp[-(x-\mu)^2/2\sigma^2]/\sigma\sqrt{2\pi}$. The mean is indicated by μ.

formula, but some of its general properties are important. The curve is symmetrical about μ and the greater the value of σ the greater the spread of the curve, as shown in Fig. 2.3. More detailed analysis shows that, whatever the values of μ and σ, approximately 68% of the population lies within $\pm 1\sigma$ of the mean, approximately 95% of values lie within $\pm 2\sigma$ of the mean and approximately 99.7% of values lie within $\pm 3\sigma$ of the mean (Fig. 2.4). This would mean that, if the nitrate ion concentrations given in Table 2.3 are normally distributed, about 68% should lie in the range 0.483–0.517, about 95% in the range 0.467–0.533 and 99.7% in the range 0.450–0.550. In fact 33 out of the 50 results (66%) lie between 0.483 and 0.517, 49 (98%) between 0.467 and 0.533, and all the results between 0.450 and 0.550, so the agreement with theory is fairly good.

Although it cannot be proved that repeated measurements of a single analytical

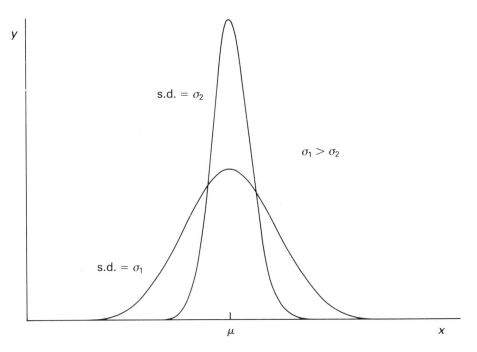

Fig. 2.3.   Normal distributions with the same mean but different values of the standard deviation.

quantity are always normally distributed, there is considerable evidence that this assumption is generally at least approximately true. Moreover, we shall see, when we come to look at sample means, that any departure of a population from normality is not usually important in the context of the statistical tests most frequently used.

The normal distribution is not only applicable to repeated measurements made on the same specimen. It also often fits the distribution of results obtained when the same quantity is measured for different materials from similar sources. For example if we measured the concentration of albumin in blood sera taken from healthy adult humans we would find the results were approximately normally distributed. However, in this second type of population, i. e. one measurement on each of a number of specimens, other distributions are not uncommon. In particular the so-called **log-normal distribution** is frequently encountered: in this distribution the *logarithms* of the concentrations (or other characteristics), when plotted against frequency, give a normal distribution curve. For example the antibody concentration in human blood sera is approximately log-normally distributed (Fig. 2.5) and the particle sizes of the droplets formed by the nebulizers used in flame spectroscopy may also follow this distribution.

In this chapter the word 'sample' has been introduced and used in its statistical sense of a group of objects selected from the population of all such objects, for example a sample of 50 measurements of nitrate ion concentration from the (infinite) population of all such possible measurements, or a sample of healthy human adults chosen from the whole population in order to measure the concentration of serum albumin for each one. The Commission on Analytical Nomenclature of the Analytical Chemistry Division of the International Union of Pure and Applied Chemistry has pointed out that confusion and ambiguity can arise if the term 'sample' is also used in its colloquial sense of the 'actual material being studied' (see bibliography). It recommends that the term sample is confined to its statistical concept. Other words should be used to describe the material on which measurements are being made, in each case preceded by 'test', for example **test solution** or **test extract**. We can then talk unambiguously of a sample of measurements on a test extract, or a sample of tablets from a batch. A test portion from a population which varies with time, such as a river or circulating blood, should be described as a **specimen**.

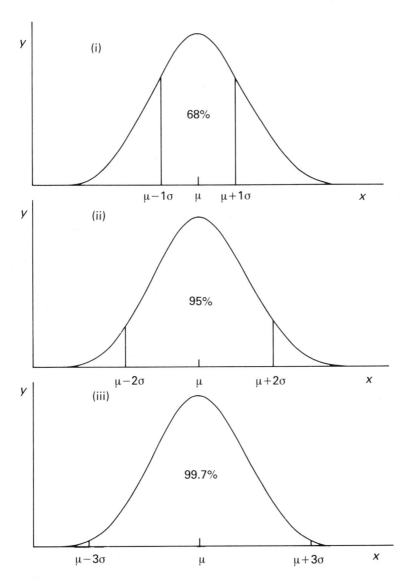

Fig. 2.4.   Properties of the normal distribution: (i) approximately 68% of values lie within ± 1σ of the mean; (ii) approximately 95% of values lie within ± 2σ of the mean; (iii) approximately 99.7% of values lie within ± 3σ of the mean.

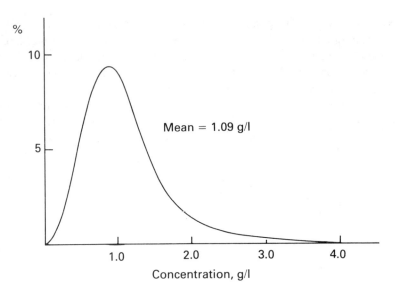

Fig. 2.5.   An approximately log-normal distribution: concentrations of serum immunoglobulin M antibody in male subjects. Note the asymmetry of the curve.

2.3   THE SAMPLING DISTRIBUTION OF THE MEAN

We have already seen that the mean of a sample of measurements provides us with an
estimate of the true value, μ, of the quantity we are trying to measure. Since,
however, the individual measurements are distributed about the true value with a
spread which depends on the precision, it is most unlikely that the mean of the
sample is *exactly* equal to the true value. For this reason it is more useful to give a
range of values within which we are almost certain the true value lies. The width of
this range depends on two factors. The first is the precision of the individual
measurements, which in turn depends on the variance of the population. The second
is the number of measurements in the sample. The very fact that we repeat
measurements implies we have more confidence in the mean of several values than in
a single one. Most people will feel that the more measurements we make, the more
reliable our estimate of μ, the true value, will be. To pursue this idea, let us return to
the nitrate ion determination described in Section 2.2. In practice it would be most
unusual to make 50 repeated measurements in such a case. A more likely number
would be 5 and we can see how the means of samples of this size are spread about μ by
treating the results in Table 2.2 as ten samples, each containing 5 results. Taking each
column as one sample, the means are 0.506, 0.504, 0.502, 0.496, 0.502, 0.492, 0.506,
0.504, 0.500, 0.486. It is at once obvious that these means are more closely clustered
than the original measurements. Just as the original measurements were a sample
from an infinite population of possible measurements, so these means are a sample
from the possible means of samples of 5 measurements from the whole population.
The distribution of these sample means is called the **sampling distribution of the
mean**. Its mean is the same as the mean of the original population. Its standard
deviation is called the **standard error of the mean (s.e.m.)**. There is an exact

mathematical relationship between it and the standard deviation, σ, of the distribution of the individual measurements, which is independent of the way in which they are distributed. If n is the sample size this relationship is:

$$\text{s.e.m.} = \sigma/\sqrt{n} \tag{2.4}$$

As we expect intuitively, the larger n is, the smaller the spread of the sample means about μ. This universally used term, the standard *error* of the mean, might mislead the reader into thinking that $\sigma/\sqrt{n}$ gives the difference between $\bar{x}$ and μ. This is not so: $\sigma/\sqrt{n}$ gives a measure of the uncertainty involved in estimating μ from $\bar{x}$, as we shall see in the next section.

Another property of the sampling distribution of the mean is that, *even if the original population is not normal*, the sampling distribution of the mean tends to the normal distribution as n increases. This result is known as the **central limit theorem**. This theorem is of the utmost importance because many statistical tests are performed on the mean and assume that it is normally distributed. Since in practice we can assume that distributions of repeated measurements are at least approximately normally distributed, it is reasonable to assume that the means of quite small samples (say > 5) are normally distributed.

2.4 CONFIDENCE LIMITS OF THE MEAN

Now that we know the form of the sampling distribution of the mean we can return to the problem of using a sample to define a range within which we may reasonably assume the true value lies. (Remember that in doing this we are assuming systematic errors to be absent.) Such a range is known as a **confidence interval** and the extreme values of the range are called the **confidence limits**. The term 'confidence' implies that we can assert with a given degree of confidence, i.e. a certain probability, that the confidence interval *does* include the true value. The size of the confidence interval will obviously depend on how certain we want to be that it includes the true value: the greater the certainty, the *greater* the interval required.

Figure 2.6 shows the sampling distribution of the mean for samples of size n. Assuming from now on that this distribution is normal, then 95% of the sample means will lie in the range given by:

$$\mu - 1.96(\sigma/\sqrt{n}) < \bar{x} < \mu + 1.96(\sigma/\sqrt{n}) \tag{2.5}$$

(The exact value 1.96 is used in this equation rather than the approximate value, 2, which is often used.)

In practice, however, we usually have one sample, of known mean, and we require a range for μ, the true value. Equation (2.5) can be rearranged to give this:

$$\bar{x} - 1.96(\sigma/\sqrt{n}) < \mu < \bar{x} + 1.96(\sigma/\sqrt{n}) \tag{2.6}$$

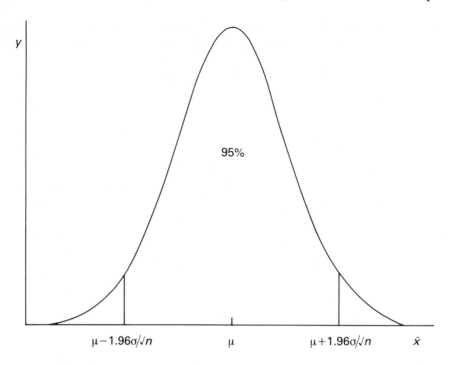

Fig. 2.6.   The sampling distribution of the mean, showing the range within which 95% of
sample means lie.

Equation (2.6) gives the **95% confidence interval** of the mean. Similarly the 99.7%
confidence interval is given by:

$$\bar{x} - 2.97(\sigma/\sqrt{n}) < \mu < \bar{x} + 2.97(\sigma/\sqrt{n}) \qquad (2.7)$$

Another confidence interval commonly used is the 99% confidence interval which is
given by:

$$\bar{x} - 2.58(\sigma/\sqrt{n}) < \mu < \bar{x} + 2.58(\sigma/\sqrt{n}) \qquad (2.8)$$

Equation (2.6) can be used to calculate the 95% confidence limits for the nitrate ion
concentration. We have $\bar{x} = 0.500$ and $n = 50$. The only quantity in the equation
which we do not know is σ. For *larger* samples such as this one, s gives a sufficiently
accurate estimate of σ to replace it. Thus the 95% confidence interval for the nitrate
ion concentration is:

$$0.500 - 1.96 \times 0.0165/\sqrt{50} < \mu < 0.500 + 1.96 \times 0.0165/\sqrt{50}$$

giving the confidence limits as:

$$\mu = 0.500 \pm 0.0046 \ \mu g/ml$$

As the sample size gets smaller, the uncertainty introduced by using s to estimate σ increases. To allow for this the equation used to calculate the confidence limits is modified to:

$$\mu = \bar{x} \pm t(s/\sqrt{n}) \tag{2.9}$$

The appropriate value of t depends both on $(n-1)$, which is known as the number of **degrees of freedom** (usually given the symbol v) and the degree of confidence required. [The term 'degrees of freedom' refers to the number of *independent* deviations $(x_i - \bar{x})$ which are used in calculating s. In this case the number is $(n-1)$, because when $(n-1)$ deviations are known the last can be deduced by using the obvious result $\sum_i (x_i - \bar{x}) = 0$]. Values of t are given in Table 2.4. It can be seen from this table that for sample sizes greater than 50 the values of t are very close to the values 1.96 and 2.58 used in Eqs. (2.6) and (2.8) respectively. This confirms the validity of the assumption used above in calculating the confidence limits for the nitrate concentration. The use of this table can be illustrated by an example.

Example. The sodium ion content of a urine specimen was determined by using an ion-selective electrode. The following values were obtained: 102, 97, 99, 98,101, 106mM. What are the 95% and 99% confidence limits for the sodium ion concentration?

The mean and standard deviation of these values are 100.5mM and 3.27mM respectively. There are 6 measurements and therefore 5 degrees of freedom. From Table 2.4 the value of t for calculating the 95% confidence limits is 2.57 and

Table 2.4.   Values of t for confidence intervals

Degrees of freedom	Values of t for confidence interval of	
	95%	99%
1	12.71	63.66
2	4.30	9.92
3	3.18	5.84
4	2.78	4.60
5	2.57	4.03
10	2.23	3.17
20	2.09	2.85
30	2.04	2.75
50	2.01	2.68
100	1.98	2.63

from Eq. (2.9) the 95% confidence limits are given by:

$$\mu = 100.5 \pm 2.57 \times 3.27/\sqrt{6}$$
$$= 100.5 \pm 3.4 mM$$

Similarly the 99% confidence limits are given by:

$$\mu = 100.5 \pm 4.03 \times 3.27/\sqrt{6}$$
$$= 100.5 \pm 5.4 \mathrm{m}M$$

2.5 PRESENTATION OF RESULTS

As has already been emphasized, no quantitative experimental result is of any value unless it is accompanied by an estimate of the errors involved in its measurement. A common practice in analytical chemistry literature is to quote the mean as the estimate of the quantity measured and the standard deviation as the estimate of the precision. Less commonly, the standard error of the mean is sometimes quoted instead of the standard deviation, or the result is given in the form of the 95% confidence limits of the mean. Since there is no universal convention it is obviously essential to state the form used and, *provided that the value of* n *is given,* the three forms can be easily interconverted by using Eqs. (2.4) and (2.9).

A related aspect of presenting results is the rounding-off of the answer. The important principle here is that the *number of significant figures given indicates the precision of the experiment.* It would clearly be absurd, for example, to give the result of a titrimetric analysis as $0.107846M$ — no analyst could achieve the implied precision of 0.000001 in ca. 0.1, i.e. 0.001%. In practice it is usual to quote as **significant figures** all the digits which are certain, plus the first uncertain one. For example, the mean of the values 10.09, 10.11, 10.09, 10.10 and 10.12 is 10.102, and their standard deviation is 0.01304. Clearly there is uncertainty in the second decimal place; the results are all 10.1 to one decimal place, but disagree in the second decimal place. Using the suggested method the result would be quoted as:

$$\bar{x} \pm s = 10.10 \pm 0.01 \ (n = 5)$$

If it was felt that this resulted in an unacceptable rounding-off of the standard deviation, then the result could be given as:

$$\bar{x} \pm s = 10.10_2 \pm 0.01_3 \ (n = 5)$$

where the use of a subscript indicates that the digit is given only to avoid loss of information. The reader could decide whether it was useful or not.

Similarly, when confidence limits are calculated [see Eq. (2.9)], there is no point in giving the value of $ts/\sqrt{n}$ to more than two significant figures. The value of $\bar{x}$ should then be given to the corresponding number of decimal places.

The number of significant figures quoted is sometimes used instead of a specific estimate of the precision of a result. For example $0.1046M$ is taken to mean that the figures in the first three decimal places are certain but there is doubt about the fourth. However, since the uncertainty in the last figure could be anything from 0.00005 to 0.0005, this method gives a poor estimate of precision and is not to be recommended. Sometimes the uncertainty in the last figure is emphasized by using the formats

0.104(6)M or 0.104_6M, but it remains preferable to give a specific estimate of precision such as the standard deviation.

One problem which arises is whether a 5 should be rounded up or down. For example, if 9.65 is rounded to one decimal place, should it become 9.6 or 9.7? It is evident that the results will be biased if a 5 is always rounded up; this bias can be avoided by rounding the 5 to the nearest *even* number giving, in this case, 9.6. Analogously, 4.75 is rounded to 4.8.

When several measured quantities are to be used to calculate a final result (see Section 2.7) these quantities should not be rounded-off too much or a needless loss of precision will result. A good rule is to keep one digit beyond the last significant figure and leave further rounding until the final result is reached. The same advice applies when the mean and standard deviation are used to apply a statistical test such as the *F*- and *t*-tests (see Chapter 3): the unrounded values of $\bar{x}$ and s should be used in the calculations.

2.6   OTHER USES OF CONFIDENCE LIMITS

Confidence intervals can be used as a test for systematic errors as shown in the following example.

Example. The absorbance scale of a spectrometer is tested at a particular wavelength with a standard solution which has an absorbance given as 0.470. Ten measurements of the absorbance with the spectrometer give $\bar{x} = 0.461$, and $s = 0.003$. Find the 95% confidence interval for the mean absorbance as measured by the spectrometer, and hence decide whether a systematic error is present.

The 95% confidence limits for the absorbance as measured by the spectrometer are [Eq. (2.9)]:

$$\mu = \bar{x} \pm t(s/\sqrt{n})$$
$$= 0.461 \pm 2.26 \times 0.003/\sqrt{10}$$
$$= 0.461 \pm 0.002$$

(The value of t was obtained from Table A.1 in the appendices, which is a more complete version of Table 2.4).

Since this confidence interval does not include the known absorbance of 0.470, it is likely that a systematic error has occurred. Another approach to the same problem is introduced in the next chapter.

Confidence limits can also be used in cases where measurements are made on each of a number of specimens. Suppose for example that the mean weight of a tablet in a very large batch is required: it would be too time-consuming to weigh each tablet. Similarly, if a destructive method of analysis such as atomic-absorption spectrometry is to be used to analyse the batch to assess mean content, it is clearly impossible to examine every tablet. In each case, a sample could be taken from the batch (which in such instances forms the population), and from the mean and standard deviation of the sample a confidence interval could be found for the mean value of the quantity measured.

2.7 PROPAGATION OF RANDOM ERRORS

In experimental work, the quantity to be determined is often calculated from a combination of observable quantities. We have already seen, for example, that even a relatively simple operation such as a titrimetric analysis involves several stages, each of which will be subject to errors (see Chapter 1). The final calculation may involve taking the sum, difference, product or quotient of two or more quantities or the raising of any quantity to a power.

It is most important to note that the procedures used for combining random and systematic errors are completely different. This is because random errors to some extent cancel each other out, whereas every systematic error occurs in a definite and known sense. Suppose for example that the final result of an experiment, x, is given by $x = a + b$. If a and b each have a systematic error of $+1$, it is clear that the systematic error in x is $+2$. If, however, a and b each have a random error of ± 1, the random error in x is *not* ± 2: this is because there will be occasions when the random error in a is positive while that in b is negative (or vice versa).

This section deals only with the propagation of random errors (systematic errors are considered in Section 2.8). If the precision of each observation is known then simple mathematical rules can be used to estimate the precision of the final result. These rules can be summarized as follows:

(i) *Linear combinations*

In this case the final value, y, is calculated from a linear combination of measured quantities a, b, c, etc. by:

$$y = k + k_a a + k_b b + k_c c + \ldots \tag{2.10}$$

where k, k_a, k_b, k_c, etc. are constants. Variance (defined as the square of the standard deviation) has the important property that the variance of a sum or difference of *independent* quantities is equal to the sum of their variances. It can be shown that if σ_a, σ_b, σ_c, etc. are the standard deviations of a, b, c, etc., the standard deviation of y, σ_y, is given by:

$$\sigma_y = \sqrt{(k_a \sigma_a)^2 + (k_b \sigma_b)^2 + (k_c \sigma_c)^2 + \ldots} \tag{2.11}$$

Example. In a titration the initial reading on the burette is 3.51 ml and the final reading is 15.67 ml, both with a standard deviation of 0.02 ml. What is the volume of titrant used and what is its standard deviation?

Volume used $= 15.67 - 3.51 = 12.16$ ml

$$\text{Standard deviation} = \sqrt{(0.02)^2 + (0.02)^2} = 0.028 \text{ ml.}$$

This example illustrates the important point that the standard deviation for the final result is larger than the standard deviations of the individual burette readings, even though the volume used is calculated from a difference, but is less than the sum of the standard deviations.

(ii) *Multiplicative expressions*
If y is calculated from an expression of the type:

$$y = kab/cd \tag{2.12}$$

(where a, b, c and d are independent measured quantities and k is a constant) then there is a relationship between the squares of the *relative* standard deviations:

$$\frac{\sigma_y}{y} = \sqrt{\left(\frac{\sigma_a}{a}\right)^2 + \left(\frac{\sigma_b}{b}\right)^2 + \left(\frac{\sigma_c}{c}\right)^2 + \left(\frac{\sigma_d}{d}\right)^2} \tag{2.13}$$

Example. The quantum yield of fluorescence, ϕ, is calculated from the expression:

$$\phi = I_f/kclI_0\,\varepsilon$$

where the quantities involved are defined below, with an estimate of their relative standard deviations in brackets:

I_0 = incident light intensity (0.5%)
I_f = fluorescence intensity (2%)
ε = molar absorptivity (1%)
c = concentration (0.2%)
l = path-length (0.2%)
k is an instrument constant.

From Eq. (2.13), the relative standard deviation of ϕ is given by:

$$\begin{aligned}
\text{r.s.d.} &= \sqrt{(2^2 + 0.2^2 + 0.2^2 + 0.5^2 + 1^2)} \\
&= \sqrt{(4 + 0.04 + 0.04 + 0.25 + 1)} \\
&= \sqrt{5.33} \\
&= 2.3\%
\end{aligned}$$

It can be seen that the relative standard deviation in the final result is not much larger than the largest relative standard deviation used to calculate it (i.e. 2% for I_f). This is mainly a consequence of the squaring of the relative standard deviations and illustrates an important general point: any efforts to improve the precision of an experiment need to be directed towards improving the precision of the least precise values. As a corollary to this, there is no point in wasting effort in increasing the precision of the most precise values. This is not to say that small errors are unimportant: small errors at many stages of an experiment, such as the titrimetric analysis discussed in detail in Chapter 1, will produce an appreciable error in the final result.

It is important to note that when a quantity is raised to a power, e.g. b^3, then the error is not calculated as for a multiplication, i.e. $b \times b \times b$, because the quantities involved are not independent. If the relationship is:

$$y = b^n \tag{2.14}$$

then the standard deviations of y and b are related by:

$$\frac{\sigma_y}{y} = \left| \frac{n\sigma_b}{b} \right| \tag{2.15}$$

(The modulus sign $|\,|$ means that the magnitude of the enclosed quantity is taken without respect to sign, e.g. $|-2| = 2$).

(iii) *Other functions*
If y is a general function of x:

$$y = f(x) \tag{2.16}$$

then the standard deviations of x and y are related by:

$$\sigma_y = \left| \sigma_x \frac{dy}{dx} \right| \tag{2.17}$$

Example. The absorbance, A, of a solution is given by:

$$A = -\log(T) \tag{2.18}$$

where T is the transmittance. If the measured value of T is 0.501 with a standard deviation of 0.001, calculate A and its standard deviation.
   We have:

$$A = -\log 0.501 = 0.300$$

Also:

$$dA/dT = -(\log e)/T = -0.434/T$$

so from Eq. (2.17):

$$\sigma_A = |\sigma_T(-\log e/T)|$$
$$= |0.001 \times (-0.434/0.501)| = 0.0008_7$$

It is interesting to note that for this widely used experimental method we can also find the conditions for which the relative standard deviation is a minimum. The relative standard deviation of A is given by

$$\text{r.s.d. of } A = 100\sigma_A/A = \frac{100\sigma_T \log e}{T \log T} \qquad (2.19)$$

Differentiation of this expression with respect to T shows that the coefficient of variation of A is a minimum when $T = 1/e = 0.368$.

Example. The Nernst equation, applied in potentiometric analysis, is

$$E = E^\circ + (RT/nF)\ln c$$

where E is the measured potential of the electrode, E° is the standard electrode potential for the ion to be determined, T is the absolute temperature, n is the number of electrons involved in the half-cell concerned, c is the concentration of the ion, R is the gas constant and F is the Faraday constant. Obtain an expression for the relative standard deviation of the concentration, assuming that $T = 298$ K and is without error. Calculate the value of the relative standard deviation if $n = 1$ and the s.d. of E is 0.001 V.

Rewriting the Nernst equation and substituting numerical values for T, R and F gives:

$$c = \exp\{40n(E - E^\circ)\}$$

Differentiating with respect to E:

$$dc/dE = 40n\exp\{40n(E - E^\circ)\}$$

so from Eq. (2.17)

$$\sigma_c = 40n\sigma_E \exp\{40n(E - E^\circ)\}$$

$$= 40nc\sigma_E$$

Thus the relative standard deviation of c is

$$100\sigma_c/c = 100 \times 40n\sigma_E\%$$

When $n = 1$ and σ_E is 0.001 V, we have

$$\text{r.s.d. of } c = 100 \times 0.001 \times 40 = 4\%$$

In contrast to the previous example, the relative standard deviation is constant.

2.8 PROPAGATION OF SYSTEMATIC ERRORS

The rules for the combination of systematic errors can also be divided into three groups:

(i) *Linear combinations*

If y is calculated from measured quantities by use of Eq. (2.10), and the systematic errors in a, b, c, etc., are Δa, Δb, Δc, etc., then the systematic error in y, Δy, is calculated from:

$$\Delta y = k_a\Delta a + k_b\Delta b + k_c\Delta c + \ldots \tag{2.20}$$

Remember that the systematic errors are either positive or negative and that these signs *must* be included in the calculation of Δy.

The total systematic error can sometimes be zero. Suppose, for example, a balance with a systematic error of -0.01 g is used for the weighings involved in making a standard solution. Since the weight of the solute used is found from the difference between two weighings the systematic errors cancel out. Carefully considered procedures, such as this, can often minimize the systematic errors, as described in Chapter 1. It should be pointed out here, however, that this applies only to an electronic balance with a single internal reference weight; when an 'old-fashioned' balance with a set of individual weights is used, the weights must be calibrated, and the appropriate corrections made (and even then, there will be a random calibration error!).

(ii) *Multiplicative expressions*

If y is calculated from the measured quantities by use of Eq. (2.12) then *relative* systematic errors are used:

$$(\Delta y/y) = (\Delta a/a) + (\Delta b/b) + (\Delta c/c) + (\Delta d/d) \tag{2.21}$$

When a quantity is raised to some power, then Eq. (2.15) is used with the modulus sign omitted and the standard deviations replaced by systematic errors.

(iii) *Other functions*

The equation used is identical to Eq. (2.17) but with the modulus sign omitted and the standard deviations replaced by systematic errors.

BIBLIOGRAPHY

O. L. Davies and P. L. Goldsmith, *Statistical Methods in Research and Production,* Longmans, London, 1982. Gives a more detailed treatment of the subject matter of this chapter.
P. Móritz, Chapter 1 in *Comprehensive Analytical Chemistry Vol. XI,* G. Svehla (ed.), Elsevier, Amsterdam, 1981. This article discusses in detail the application of the theory of errors to analytical methods.
D. A. Skoog and D. M. West, *Fundamentals of Analytical Chemistry,* 4th Ed., Holt Saunders, New York, 1982. Describes the use of statistics in evaluating analytical data.
J. Topping, *Errors of Observation and their Treatment,* Chapman & Hall, London, 1962. Gives a fuller treatment of the theory of errors and discusses the theory that errors are normally distributed.
Nomenclature for Sampling in Analytical Chemistry, *Pure and Applied Chemistry,* 1990, **62,**1193.

EXERCISES

(1) To investigate the reproducibility of a method for the determination of selenium in foods, nine measurements were made on a single batch of brown rice with the following results:

Sample	Selenium, μg/g
1	0.07
2	0.07
3	0.08
4	0.07
5	0.07
6	0.08
7	0.08
8	0.09
9	0.08

(T. Moreno-Domínguez, C. García-Moreno and A. Mariné-Font, *Analyst,* 1983, **108,** 505).
Calculate the mean, standard deviation and relative standard deviation of these results.

(2) Seven measurements of the pH of a buffer solution gave the following results:

5.12, 5.20, 5.15, 5.17, 5.16, 5.19, 5.15.

Calculate (i) the 95% and (ii) the 99% confidence limits for the true pH. (Assume that there are no systematic errors.)

(3) Ten analyses of the concentration of albumin gave a mean of 20.92 g/l. and a standard deviation of 0.45 g/l. (J. W. Foote and H. T. Delves, *Analyst,* 1983, **108,** 492).
Calculate the 95% confidence limits of the mean.

(4) The concentration of lead in the blood stream was measured for a sample of 50 children from a large school near a busy main road. The sample mean was 10.12 ng/ml and the standard deviation was 0.64 ng/ml. Calculate the 95% confidence interval for the mean lead concentration for all the children in the school.
About how big should the sample have been to reduce the range of the confidence interval to 0.2 ng/ml (i.e. ± 0.1 ng/ml)?

(5) In an investigation into the accuracy and precision of a method for determining arprinocid in feed premixes, six replicate determinations were made on a premix formulated to contain 10.2% arprinocid. The results were:

10.4, 10.4, 10.6, 10.3, 10.5, 10.5%.

(J. D. Stong and D. W. Fink, *Analyst*, 1982, **107,** 113).
Calculate the mean, standard deviation and 95% and 99% confidence limits of the mean. Is the formulated value of 10.2% within (i) the 95% and (ii) the 99% confidence limits?

(6) Ten measurements of the ratio of two peak areas in a liquid chromatogrpahy experiment gave the following values:

0.2911, 0.2898, 0.2923, 0.3019, 0.2997, 0.2961, 0.2947, 0.2986, 0.2902, 0.2882.

(P. Jonvel and G. Andermann, *Analyst*, 1983, **108,** 411).
Calculate the mean, standard deviation and 99% confidence limits of the mean.

(7) A $0.1M$ solution of acid was used to titrate 10 ml of $0.1M$ solution of alkali and the following volumes of acid were recorded:

9.88, 10.18, 10.23, 10.39, 10.25 ml.

Calculate the 95% confidence limits of the mean and use them to decide if there is any evidence of systematic error.

(8) This problem considers the random errors involved in making up a standard solution. A volume of 250 ml of a $0.05M$ solution of a reagent of formula weight (relative molecular mass) 40 was made up, the weighing being done by difference. The standard deviation of each weighing was 0.0001 g: what were the standard deviation and relative standard deviation of the weight of reagent used? The standard deviation of the volume of solvent used was 0.05 ml. Express this as a relative standard deviation. Hence calculate the relative standard deviation of the molarity of the solution.
Repeat the calculation for a reagent of formula weight 392.

(9) The solubility product of barium sulphate is 1.3×10^{-10}, with a standard deviation of 0.1×10^{-10}. Calculate the standard deviation of the calculated solubility of barium sulphate in water.

3

Significance tests

3.1 INTRODUCTION

One of the most important properties of an analytical method is that it should be free from systematic error, i.e. the value which it gives for the amount of the analyte should be the *true* value. This property may be tested by applying the method to a standard sample containing a known amount of analyte (Chapter 1). However, as we saw in the last chapter, random errors make it most unlikely that the measured amount would *exactly* equal the standard amount even if there were no systematic error. In order to decide if the difference between the measured and standard amounts can be accounted for by these random errors a statistical test known as a **significance test** can be employed. As its name implies, this approach tests whether the difference between the two results is significant, or can be accounted for merely by random variations. Significance tests are widely used in the evaluation of experimental results. This chapter considers several tests which are particularly useful to analytical chemists.

3.2 COMPARISON OF AN EXPERIMENTAL MEAN WITH A KNOWN VALUE

In making a significance test we are testing the truth of a hypothesis which is known as a **null hypothesis**. For the example in the previous paragraph we adopt the null hypothesis that the analytical method is *not* subject to systematic error. The term *null* is used to imply that there is *no* difference between the observed and known values other than that which can be attributed to random variation. Assuming that this null hypothesis is true, statistical theory can be used to calculate the probability that the observed difference (or a greater one) between the sample mean, $\bar{x}$, and the true value, μ, arises solely as a result of random errors. The lower the probability that the observed difference occurs by chance, the less likely it is that the null hypothesis is true. Usually the null hypothesis is rejected if the probability of such a difference occurring by chance is *less* than 1 in 20 (i.e. 0.05 or 5%) and in such a case the difference is said to be **significant at the 0.05 (or 5%) level**. Using this level of significance there is, on average, a 1 in 20 chance that we shall reject the null hypothesis *when it is in fact true*. In order to be more certain that we make the correct decision a higher level of significance can be used, usually 0.01 or 0.001 (1% or 0.1%). The significance level is indicated by writing, for example, P (i.e. probability) = 0.05, and gives the probability of rejecting a true null hypothesis. It is important to appreciate that if the null hypothesis is retained it has not been *proved* that it is true, only that it has not been demonstrated to be false. Later in the chapter the probability of retaining a null hypothesis *when it is in fact false* will be discussed.

In order to decide whether the difference between μ and $\bar{x}$ is significant, Eq. (2.9):

$$\mu = \bar{x} \pm (ts/\sqrt{n})$$

(where n is the sample size) is rewritten:

$$t = (\bar{x} - \mu)\sqrt{n}/s \tag{3.1}$$

and a value of t is calculated by substituting the experimental results in this equation. If $|t|$ (i.e. the calculated value of t without regard to sign) *exceeds* a certain **critical value** then the null hypothesis is rejected. The critical value of $|t|$ for a particular significance level is found from Table A.1. For example for a sample size of 10 (i.e. 9 degrees of freedom) and a significance level of 0.01, the critical value of $|t|$ is 3.25.

> *Example.* In a method for determining mercury by cold-vapour atomic absorption the following values were obtained for a standard reference material containing 38.9% mercury:
>
> 38.9, 37.4, 37.1%
>
> (P.-K. Hou, O.-W. Lau and M.-C. Wong, *Analyst*, 1983, **108**, 64).
> Is there any evidence of systematic error?
>
> The mean of these values is 37.8% and the standard deviation is 0.964%. Adopting the null hypothesis that there is no systematic error, i.e. $\mu = 38.9\%$, and using Eq. (3.1) gives
>
> $$|t| = |(37.8 - 38.9) \times \sqrt{3}/0.964| = 1.98$$

From Table A.1, for 2 degrees of freedom the critical value of $|t|$ is 4.3 ($P = 0.05$). Since the observed value of $|t|$ is less than the critical value the null hypothesis is retained: there is no evidence of systematic error. Note again that this does not mean that there *are* no systematic errors, only that they have not been demonstrated.

3.3 COMPARISON OF THE MEANS OF TWO SAMPLES

Another way in which the results of a new analytical method may be tested is by comparing them with those obtained by using a second (perhaps a reference) method. In this case we have two sample means $\bar{x}_1$ and $\bar{x}_2$. Taking the null hypothesis that the two methods give the same result, we need to test whether $(\bar{x}_1 - \bar{x}_2)$ differs significantly from zero. If the two samples have standard deviations which are not significantly different (see Section 3.5 for a method of testing this assumption), a **pooled** estimate of standard deviation can be calculated from the two individual standard deviations s_1 and s_2 by using the equation:

$$s^2 = \{(n_1 - 1)s_1^2 + (n_2 - 1)s_2^2\}/(n_1 + n_2 - 2) \tag{3.2}$$

It can be shown that t is then given by:

$$t = (\bar{x}_1 - \bar{x}_2)/s\sqrt{(1/n_1 + 1/n_2)} \tag{3.3}$$

where t has $n_1 + n_2 - 2$ degrees of freedom.

Example. In a comparison of two methods for the determination of boron in plant samples the following results (μg/g) were obtained:

spectrophotometric method: mean = 28.0; standard deviation = 0.3
fluorimetric method: mean = 26.25; standard deviation = 0.23

For each method 10 determinations were made.
(J.Anarez, A.Bonilla and J.C.Vidal, *Analyst*, 1983, **108**, 368).

Do these two methods give results having means which differ significantly?
The null hypothesis adopted is that the means of the results given by the two methods are equal. From Eq. (3.2), the pooled value of the standard deviation is given by:

$$s^2 = (9 \times 0.3^2 + 9 \times 0.23^2)/18$$
$$= 0.0715$$
$$s = 0.267$$

From Eq. (3.3):

$$t = (28.0 - 26.25)/0.267\sqrt{(1/10 + 1/10)}$$
$$= 14.7$$

There are 18 degrees of freedom, so (Table A.1) the critical value of $|t|$

$(P = 0.05)$ is 2.1: since the experimental value of $|t|$ is greater than this the difference between the two results is significant at the 5% level and the null hypothesis is rejected. In fact since the critical value of $|t|$ for $P = 0.001$ is about 3.9, the difference is significant even at the 0.1% level. In other words, if the null hypothesis is true the probability of such a large difference arising by chance is less than 1 in 1000.

Another application of this test is illustrated by the following example where it is used to decide whether a change in the conditions of an experiment affects the result.

Example. In a series of experiments on the determination of tin in foodstuffs, samples were boiled with hydrochloric acid under reflux for different times. Some of the results are shown below:

Refluxing time (min)	Tin found (mg/kg)
30	55, 57, 59, 56, 56, 59
75	57, 55, 58, 59, 59, 59

(Analytical Methods Committee, *Analyst*, 1983, **108**, 109).

Does the mean amount of tin found differ significantly for the two boiling times?
    The mean and variance (square of the standard deviation) for the two times are:

$$30 \text{ min} \quad \bar{x}_1 = 57.00 \quad s_1^2 = 2.80$$

$$75 \text{ min} \quad \bar{x}_2 = 57.83 \quad s_2^2 = 2.57$$

The null hypothesis is adopted that boiling has no effect on the amount of tin found. By Eq. (3.2), the pooled value for the variance is given by:

$$s^2 = (5 \times 2.80 + 5 \times 2.57)/10$$

$$= 2.685$$

$$s = 1.64$$

t is calculated from Eq. (3.3):

$$t = (57.00 - 57.83)/1.64 \sqrt{(1/6 + 1/6)}$$

$$= -0.88$$

There are 10 degrees of freedom so the critical value of $|t|$ is 2.23 ($P = 0.05$). The observed value of $|t|$ is less than the critical value so the null hypothesis is retained: there is no evidence that the length of boiling time affects the recovery rate.

The use in significance testing of critical values from statistical tables was adopted because it was formerly too tedious to calculate the probability of t exceeding the experimental value. Modern computers have altered this situation, and commercial significance testing software often quotes the results of this and other tests in terms of a probability. Thus, for the example above, a computer program gave the result as $P (|t| \geqslant 0.88) = 0.40$. Since this probability is much *greater* than 0.05, the result is not significant at $P = 0.05$, in agreement with the previous calculation. Obviously the power to calculate an exact probability is a great advantage, removing the need for statistical tables containing critical values. However, the examples in this book use critical values, as not all readers may have access to suitable software, and many scientists continue to perform significance tests with the aid of hand-held calculators, which do not normally provide P values.

If the assumption that the population standard deviations are equal is not valid, an approximate method is to modify Eq. (3.3) to:

$$t = (\bar{x}_1 - \bar{x}_2)/\sqrt{(s_1^2/n_1 + s_2^2/n_2)} \tag{3.4}$$

and calculate the number of degrees of freedom from:

$$\text{degrees of freedom} = \left(\frac{(s_1^2/n_1 + s_2^2/n_2)^2}{\dfrac{(s_1^2/n_1)^2}{n_1 + 1} + \dfrac{(s_2^2/n_2)^2}{n_2 + 1}} \right) - 2 \tag{3.5}$$

the result being rounded to the nearest whole number.

Example. The following table gives the concentration of thiol in the lysate of the blood of two groups of volunteers, the first group being 'normal' and the second suffering from rheumatoid arthritis:

Thiol concentration (mM)

Normal	Rheumatoid
1.84	2.81
1.92	4.06
1.94	3.62
1.92	3.27
1.85	3.27
1.91	3.76
2.07	

(J. C. Banford, D. H. Brown, A. A. McConnell, C. J. McNeil, W. E. Smith, R. A. Hazelton and R. D. Sturrock, *Analyst*, 1983, **107**, 195).

From these data we have:

$$n_1 = 7, \quad \bar{x}_1 = 1.921, \quad s_1 = 0.076$$
$$n_2 = 6, \quad \bar{x}_2 = 3.465, \quad s_2 = 0.440$$

Again the null hypothesis is adopted that the mean concentration of thiol is the same for the two groups. Substitution in Eq. (3.4) gives $t = 8.5$ and from Eq. (3.5) there are 5 degrees of freedom. The critical value of $|t|$ ($P = 0.01$) is 4.03 so the null hypothesis is rejected: the mean concentration of thiol is different for the two groups.

3.4 PAIRED *t*-TEST

It frequently happens that two methods of analysis have to be compared by studying test samples containing substantially different amounts of analyte, as illustrated by the following example:

Example. The following table gives the concentration of lead (μg/l.) determined by two different methods for each of four test portions:

Portion	Wet oxidation	Direct extraction
1	71	76
2	61	68
3	50	48
4	60	57

(B. M. Smith and M. B. Griffiths, *Analyst*, 1982, **107**, 253).

Do the two methods give values for the mean lead concentration which differ significantly?

The test for comparing two means (Section 3.3) is not appropriate in this case because any variation due to method will be swamped by the effect of the differences between the test portions. This difficulty is overcome by looking at the difference between each pair of results given by the two methods. Adopting the null hypothesis that there is no significant difference in the mean concentrations given by the two methods, we can test whether the mean of the differences differs significantly from zero.

For the pairs of values above, the differences are -5, -7, 2, 3. The mean difference, $\bar{x}_d$, is -1.75 and the standard deviation of the differences, s_d, is 4.99. Since $\mu_d = 0$, Eq. (3.1) for calculating t becomes:

$$t = \bar{x}_d \sqrt{n}/s_d \tag{3.6}$$

where t has $(n-1)$ degrees of freedom. Substituting in Eq. (3.6) gives $t = -0.70$. The critical value of $|t|$ is 3.18 ($P = 0.05$) and since the calculated value of $|t|$ is less than this the null hypothesis is retained: the methods do not give significantly different values for the mean lead concentration.

Alternatively, use of a computer program yielded the result $P(|t| \geqslant 0.70) = 0.53$. Since this probability is greater than 0.05 we reach the same conclusion: the two methods do not differ significantly at $P = 0.05$.

There are various circumstances in which it may be necessary or desirable to design an experiment so that each sample is analysed by each of two methods and the results are thus naturally paired. Some examples are:

(a) the quantity of any one test sample is sufficient for only one determination by each method;

(b) the methods are to be compared by using a wide variety of samples from different sources and possibly with very different concentrations (but see the next paragraph);

(c) the test samples may be presented over an extended period so it is necessary to remove the effects of variations in the environmental conditions such as temperature, pressure etc.

As analytical methods usually have to be applicable over a wide range of concentrations, a new method is often compared with a standard method by analysis of samples in which the analyte concentration may vary over several powers of ten. In this case it is inappropriate to use the paired t-test since its validity rests on the assumption that any errors, either random or systematic, are independent of concentration. Over wide ranges of concentration this assumption may no longer be true. The preferred statistical method in such cases is linear regression: its application to this situation is described later (Section 5.9).

3.5 ONE-TAILED AND TWO-TAILED TESTS

The methods described so far in this chapter have been concerned with testing the difference between two means *in either direction*. For example, the method described in Section 3.2 tests whether there is a significant difference between the experimental result and the known value for the reference material, regardless of the sign of the difference. In most situations of this kind the analyst has no preconceived idea, *prior to the experimental measurements*, as to whether any significant difference between the experimental mean and the reference value will be positive or negative. He thus needs a test that will cover either possibility: such a test is called **two-tailed** (or **two-sided**). In a few cases, however, a different kind of test may be appropriate. Consider, for example, an experiment in which it is hoped to *increase* the rate of reaction by addition of a catalyst. In this case, it is clear *before the experiment begins* that the only result of interest is whether the new rate is *greater* than the old, and only an increase need be tested for significance. This kind of test is called **one-tailed** (or **one-sided**). For a given value of n and a particular probability level, the critical value for a one-tailed test differs from that for a two-tailed test. In a one-tailed test for an increase the critical value of t (rather than $|t|$) for $P=0.05$ is that value which is *exceeded* with a probability of 5%. Since the sampling distribution of the mean is assumed to be symmetrical, this probability is half the probability that is important in the two-tailed test. The appropriate value for the one-tailed test is thus found in the $P=0.10$ column of Table A.1. Similarly, for a one-tailed test with $P=0.01$, the value in the $P=0.02$ column is used. For a one-tailed test for a *decrease*, the critical value of t will be of equal magnitude but with a negative sign.

Example. It is suspected that an acid–base titrimetric method has a significant indicator error and thus tends to give results with a positive systematic error (i.e. positive bias). To test this an exactly $0.1M$ solution of acid is used to titrate 25.00 ml of an exactly $0.1M$ solution of alkali, with the following results (ml):

$$25.06 \quad 25.18 \quad 24.87 \quad 25.51 \quad 25.34 \quad 25.41$$

Test for positive bias in these results.

From these data we have:

mean = 25.228 ml; standard deviation = 0.238 ml

Adopting the null hypothesis that there is no bias, i.e. $\mu=25.00$, and using Eq. (3.1) gives

$$t = (25.228 - 25.00) \times \sqrt{6}/0.238 = 2.35$$

From Table A.1 the critical value of t for 5 degrees of freedom is 2.02 ($P=0.05$, one-tailed test). Since the observed value of t is greater than this the null hypothesis is rejected and there is evidence for positive bias.

It is interesting to note that if a two-tailed test had been made in the example above ($|t|=2.57$) the null hypothesis would *not* have been rejected! This apparently contradictory result is explained by the fact that the decision on whether to make a one- or two-tailed test depends on the degree of prior knowledge, in this case a suspicion or expectation of positive bias. Obviously it is essential that the decision on whether the test is one- or two-tailed should be made *before* the experiment has been done, and not with hindsight, when the results might prejudice the choice. In general, it will be found that two-tailed tests are much more commonly used than one-tailed ones, and that the relatively rare circumstances in which one-tailed tests are necessary are easily identified.

3.6   *F*-TEST FOR THE COMPARISON OF STANDARD DEVIATIONS

The significance tests so far described in this chapter are used for comparing means, and hence for detecting systematic errors. It is also important in many cases to compare the standard deviations, i.e. the random errors of two sets of data. As with tests on means, this comparison can take two forms. Either we may wish to test whether Method A is more precise than Method B (i.e. a one-tailed test) or we may wish to test whether Methods A and B differ in their precision (i.e. a two-tailed test). Thus, if we wished to test whether a new analytical method is *more* precise than a standard method we would use a one-tailed test: if we wished to test whether two standard deviations differ significantly (e.g. before applying a *t*-test — see Section 3.3 above) a two-tailed test is appropriate.

The **F-test** considers the ratio of the two sample variances, i.e. the ratio of the squares of the standard deviations. The quantity calculated (F) is given by:

$$F = s_1^2/s_2^2 \qquad (3.7)$$

s_1^2 and s_2^2 being allocated in the equation so that F is always ≥ 1. The null hypothesis adopted is that the populations from which the samples are taken are normal, and that the population variances are equal. If the null hypothesis is true then the variance ratio should be close to 1. Differences from 1 occur because of random variation but if the difference is too great it can no longer be attributed to this cause: if the calculated value of F exceeds a certain critical value (obtained from tables) then the null hypothesis is rejected. This critical value of F depends on the size of both the samples, the significance level and the type of test performed. The values for $P = 0.05$ are given in Tables A.2 and A.3; the use of the Tables is illustrated in the following examples.

Example. A proposed method for the determination of the chemical oxygen demand of wastewater was compared with the standard (mercury salt) method. The following results were obtained for a sewage effluent sample:

	Mean (mg/l.)	Standard deviation (mg/l.)
Standard method	72	3.31
Proposed method	72	1.51

For each method 8 determinations were made.

(D. Ballinger, A. Lloyd and A. Morrish, *Analyst*, 1982, **107**, 1047)

Is the precision of the proposed method significantly greater than that of the standard method?

We have to decide whether the variance of the standard method is significantly greater than that of the proposed method. F is given by the ratio of the variances:

$$F_{7,7} = 3.31^2/1.51^2 = 4.8$$

The samples both comprise 8 values so the number of degrees of freedom in each case is 7, as indicated by the subscripts. This is a case where a one-tailed test must be used, the only point of interest being whether the proposed method is *more* precise than the standard method. In Table A.2 the number of degrees of freedom of the denominator is given in the left-hand column and the number of degrees of freedom of the numerator at the top. The critical value of F in this case is 3.787 ($P=0.05$). Since the calculated value of F (4.8) exceeds this, the variance of the standard method *is* significantly greater than that of the proposed method at the 5% probability level, i.e. the proposed method is more precise.

Example. In the example at the beginning of Section 3.3 it was assumed that the variances of the two methods for determining boron in plants did not differ significantly. This assumption can now be tested. The standard deviations were 0.3 and 0.23 (each obtained from 10 measurements on a specimen of a particular plant). Calculating F so that it is greater than 1, we have:

$$F_{9,9} = 0.3^2/0.23^2 = 1.7$$

In this case, however, we have no reason to expect in advance that the variance of one method should be greater than the other, so a two-tailed test is appropriate. The critical values given in Table A.2 are the values that F exceeds with a probability of 0.05 assuming that it *must* be greater than 1. In a two-tailed test the ratio of the first to the second variance could be less or greater than 1 but, if F is calculated so that it *is* greater than 1, the probability that it exceeds the critical values given in Table A.2 will be doubled. Thus these critical values are not appropriate for a two-tailed test and Table A.3 is used instead. From this Table, taking the number of degrees of freedom of both numerator and denominator as 9, the critical value of F is 4.026. The calculated value is less than this, so there is no significant difference between the two variances at the 5% level.

As with the t-test, other significance levels may be used for the F-test and the critical values can be found from the tables listed in the bibliography at the end of Chapter 1. Care must be taken that the correct table is used depending on whether the test is one- or two-sided: for an α% significance level the α% points of the F-distribution are used for a one-sided test and the $\frac{1}{2}\alpha$% points are used for a two-sided test.

3.7 OUTLIERS

Every experimentalist is familiar with the situation in which one (or possibly more) of a set of results appears to differ unreasonably from the others in the set. Such a measurement is called an **outlier**. In some cases an outlier may be attributed to a human error. For example if the following results were given for a titration:

12.12, 12.15, 12.13, 13.14, 12.12 ml

then the fourth value is almost certainly due to a slip in writing down the result and should read 12.14. However, even when such obviously erroneous values have been removed or corrected, values which appear to be outliers may still occur. Should they be kept, come what may, or should some means be found to test statistically whether or not they should be rejected? Obviously the final values presented for the mean and standard deviation will depend on whether or not the outliers are rejected. Since discussion of the precision and accuracy of a method depends on these final values, it should always be made clear whether outliers have been rejected, and if so, why.

One way (amongst several different tests) of assessing a suspect measurement is to compare the difference between it and the measurement nearest to it in size, with the difference between the highest and lowest measurements. The ratio of these differences (without regard to sign) is known as **Dixon's Q**.

$$Q = |\text{suspect value} - \text{nearest value}|/(\text{largest value} - \text{smallest value}) \qquad (3.8)$$

The critical values of Q for $P = 0.05$ and $P = 0.01$ are given in Table A.4. If the calculated value of Q exceeds the critical value the suspect value is rejected. The values given are for a two-sided test, which is appropriate when it is not known at which extreme an outlier may occur.

Example. The following values were obtained for the nitrite concentration (mg/l.) in a sample of river water:

0.403, 0.410, 0.401, 0.380

The last measurement is suspect: should it be rejected?
    We have:

$$Q = |0.380-0.401|/(0.410-0.380) = 0.021/0.03 = 0.7$$

From Table A.4, for sample size 4, the critical value of Q is 0.831 ($P=0.05$). Since the calculated value of Q does not exceed this the suspect measurement should be retained.

Ideally, further measurements should be made when a suspect value occurs, particularly if only a few values have been obtained initially. This may make it clearer whether or not the suspect value should be rejected, and if it is still retained will also reduce to some extent its effect on the mean and standard deviation.

Example. If three further measurements were added to those given in the example above so that the complete results became:

0.403, 0.410, 0.401, 0.380, 0.400, 0.413, 0.411

should 0.380 still be retained?

The calculated value of Q is now

$$Q = |0.380-0.400|/(0.413-0.380) = 0.606$$

The critical value of Q ($P=0.05$) for a sample size 7 is 0.570 so the suspect measurement is rejected at the 5% significance level.

It is important to appreciate that for a significance level of 5% there is still a chance of 5%, or 1 in 20, of incorrectly rejecting the suspect value. This may have a considerable effect on the estimation of the precision of an experiment. For example, for all 7 values of the nitrite concentration given above, the standard deviation is 0.011 mg/l., but when the suspect value is rejected the standard deviation becomes 0.0056 mg/l., i.e. the precision appears to have improved by a factor of 2.

The example above illustrates the importance of caution in rejecting outliers. When measurements are repeated only a few times (which is common in analytical work), rejection of one value makes a great difference to the mean and standard deviation. In particular, the practice of making three measurements and rejecting the one which differs most from the other two should be avoided. It can be shown that a more reliable estimate of the mean is obtained, on average, by using the middle one of the three values rather than the mean of the two unrejected values.

If a set of data contains two or more suspect results, further complications arise in deciding whether rejection is justified. Fig. 3.1 illustrates in the form of 'dot plots' two examples of such difficulties. In Fig. 3.1(a) there are two results (2.9, 3.1), both of which are suspiciously high compared with the rest of the data, yet if Q were calculated uncritically using Eq. (3.8) we would obtain:

$$Q = (3.1 - 2.9) / (3.1 - 2.0) = 0.18$$

a value which is not significant ($P = 0.05$). Clearly the possible outlier 3.1 has been **masked** by the other possible outlier, 2.9, giving a low value of Q. How are such situations tackled? Space does not permit a full discussion of this problem (see the bibliography at the end of the chapter), but it is clear that two approaches are possible. One method (a **block procedure**) would consider the two suspect results as a pair, with the test statistic being (for example) their mean divided by the mean of the whole data set. Such methods carry some risk that both suspect results are

Fig 3.1.   Dot-plots illustrating the problem of handling outliers (a) when there are two suspect results at the high end of the sample data, and (b) when there are two suspect results, one at each extreme of the data.

(necessarily) rejected or accepted together in situations where one result should really be accepted and the other rejected. Alternatively, in a **consecutive procedure**, we would first test, with the aid of a statistic similar to Q, to see whether the value 2.9 can be rejected: if it can, then naturally the value 3.1 will be rejected as well. If 2.9 is retained, a separate test is applied to the value 3.1. Considerable research is still being carried out on the relative merits of these block and consecutive methods.

A different situation arises in Fig. 3.1(b), where the two suspect values are clearly at opposite ends of the data set. Again several test methods suggest themselves, one possible statistic being $(x_n - x_1)/s$, where x_1 and x_n are the two suspect results, and s the standard deviation of the sample.

Our discussions of errors have so far assumed that the distribution of repeated measurements is normal or nearly so. It is important to realize that the outlier tests described above should be used with this assumption firmly in mind: a result that seems to be an outlier on the assumption of a normal population distribution may well not be an outlier if the sample actually comes from (for example) a log-normal distribution (Section 2.2). Therefore outlier tests should not be used if there is a doubt about the parent population distribution. This difficulty, along with the extra complications arising in cases of multiple outliers, explains the increasing use of the non-parametric and robust statistical methods described in Chapter 6. Such methods are either insensitive to extreme values, or at least give them less weight in calculations, so the problem of whether or not to reject outliers is avoided.

3.8 ANALYSIS OF VARIANCE

In Section 3.3 a method was described for comparing two means to test whether they differ significantly. In analytical work there are often more than two means to be compared. Some possible situations are: comparing the mean concentration of protein in solution, for samples stored under different conditions; comparing the mean results obtained for the concentration of an analyte by several different methods; comparing the mean titration results obtained by several different experimentalists using the same apparatus. In all these examples there are two possible sources of variation. The first, which is always present, is due to the random error in measurement. This was discussed in detail in the previous chapter: it is this error which causes a different result to be obtained each time a measurement is repeated under the same conditions. The second possible source of variation is due to what is known as a **controlled** or **fixed-effect factor**: for the examples above, the controlled factors are, respectively, the conditions under which the solution was stored, the method of analysis used and the experimentalist carrying out the titration. **Analysis of variance** (frequently abbreviated to **ANOVA**) is an extremely powerful statistical technique which can be used to separate and estimate the different causes of variation. For the particular examples above, it can be used to separate any variation which is caused by changing the controlled factor, from the variation due to random error. It can thus test whether altering the controlled factor leads to a significant difference between the mean values obtained.

ANOVA can also be used in situations where there is more than one source of random variation. Consider, for example, the purity testing of a barrelful of sodium chloride. Samples are taken from different parts of the barrel chosen at random and replicate analyses performed on these samples. In addition to the random error in the measurement of the purity, there may also be variation in the purity of the samples from different parts of the barrel. Since the samples were chosen at random, this variation will be random and is thus sometimes known as a **random-effect factor.** Again, ANOVA can be used to separate and estimate the sources of variation.

Both types of statistical analysis described above, i.e. where there is *one* factor, either controlled or random, in addition to the random error in measurement, are known as **one-way** ANOVA. The arithmetical procedures are similar in the fixed- and random-effect factor cases: examples of the former are given in this chapter and of the latter in the next chapter, where sampling is considered in more detail. More complex situations in which there are two or more factors, possibly interacting with each other, are considered in Chapter 7 (Experimental Design).

3.9 COMPARISON OF SEVERAL MEANS

Table 3.1 shows the results obtained in an investigation into the stability of a fluorescent reagent stored under different conditions. The values given are the fluorescence signals (in arbitrary units) from dilute solutions of equal concentration. Three replicate measurements were made on each sample. The table shows that the mean values for the four samples are different. However, we know that because of random error, even if the true value which we are trying to measure is unchanged, the sample mean may vary from one sample to the next. ANOVA tests whether the difference between the sample means is too great to be explained by the random error.

Table 3.1.　Fluorescence signal from solutions stored under different conditions

Conditions	Replicate measurements	Mean
A　Freshly prepared	102, 100, 101	101
B　Stored for 1 hr in the dark	101, 101, 104	102
C　Stored for 1 hr in subdued light	97, 95, 99	97
D　Stored for 1 hr in bright light	90, 92, 94	92
	Overall mean	98

Table 3.2.　Generalization of Table 3.1

						Mean
Sample 1	x_{11}	x_{12}	...... x_{1j}	......	x_{1n}	$\bar{x}_1$
Sample 2	x_{21}	x_{22}	...... x_{2j}	......	x_{2n}	$\bar{x}_2$
	⋮	⋮	⋮		⋮	⋮
Sample i	x_{i1}	x_{i2}	...... x_{ij}	......	x_{in}	$\bar{x}_i$
	⋮	⋮	⋮		⋮	⋮
Sample h	x_{h1}	x_{h2}	...... x_{hj}	......	x_{hn}	$\bar{x}_h$
				overall mean	=	$\bar{x}$

The problem can be generalised to consider h samples each with n members as in Table 3.2 where x_{ij} is the jth measurement of the ith sample. The means of the samples are $\bar{x}_1, \bar{x}_2, \ldots, \bar{x}_n$ and the mean of all the values grouped together is $\bar{x}$. The null hypothesis adopted is that all the samples are drawn from a population with mean μ and variance σ_0^2. On the basis of this hypothesis σ_0^2 can be estimated in two ways, one involving the variation *within* the samples and the other the variation *between* the samples.

(i) *Within-sample variation*

For each sample a variance can be calculated by using the formula $\Sigma(x_i - \bar{x})^2/(n-1)$ [see Eq. (2.2)]. Using the values in Table 3.1 we have:

$$\text{Variance of sample A} = \frac{(102 - 101)^2 + (100 - 101)^2 + (101 - 101)^2}{3 - 1}$$

$$= 1$$

$$\text{Variance of sample B} = \frac{(101 - 102)^2 + (101 - 102)^2 + (104 - 102)^2}{3 - 1}$$

$$= 3$$

$$\text{Variance of sample C} = \frac{(97 - 97)^2 + (95 - 97)^2 + (99 - 97)^2}{3 - 1}$$

$$= 4$$

$$\text{Variance of sample D} = \frac{(90 - 92)^2 + (92 - 92)^2 + (94 - 92)^2}{3 - 1}$$

$$= 4$$

Averaging these values gives:

$$\text{within-sample estimate of } \sigma_0^2 = (1 + 3 + 4 + 4)/4 = 3$$

This estimate has 8 degrees of freedom: each sample estimate has 2 degrees of freedom and there are 4 samples. Note that this estimate does not depend on the means of the samples: for example, if all the measurements for sample A were increased by say, 4, this estimate of σ_0^2 would be unaltered.

The general formula for the within-sample estimate of σ_0^2 is:

$$\text{within-sample estimate of } \sigma_0^2 = \Sigma_i \Sigma_j (x_{ij} - \bar{x}_i)^2/h(n-1) \tag{3.9}$$

The summation over j and division by $(n-1)$ gives the variance of each sample; the summation over i and division by h averages these sample variances. The expression in Eq. (3.9) is known as a **mean square** since it involves a sum of squared terms divided by the number of degrees of freedom. Since in this case the number of degrees of freedom is 8 and the mean square is 3, the sum of the squared terms is $3 \times 8 = 24$.

(ii) *Between-sample variation*

If the samples are all drawn from a population which has variance σ_0^2, then their means come from a population with variance σ_0^2/n (cf. the sampling distribution of the mean, Section 2.3). Thus, if the null hypothesis is true, the variance of the means of the samples gives an estimate of σ_0^2/n. From Table 3.1:

$$\text{sample mean variance} = \frac{(101-98)^2 + (102-98)^2 + (97-98)^2 + (92-98)^2}{4-1}$$

$$= \frac{62}{3}$$

so

$$\text{between-sample estimate of } \sigma_0^2 = \frac{62}{3} \times 3 = 62$$

This estimate has 3 degrees of freedom since it is calculated from 4 sample means. Note that this estimate of σ_0^2 does not depend on the variability *within* each sample, since it is calculated from the sample means. However, if, for example, the mean of sample D were changed, then this estimate of σ_0^2 would also be changed.

In general we have:

$$\text{between-sample estimate of } \sigma_0^2 = n\sum_i (\bar{x}_i - \bar{x})^2/(h-1) \tag{3.10}$$

which is again a 'mean square' involving a sum of squared terms divided by the number of degrees of freedom. In this case the number of degrees of freedom is 3 and the mean square is 62, so the sum of the squared terms is $3 \times 62 = 186$.

Summarizing our calculations so far:

$$\text{within-sample mean square} = 3 \text{ with 8 d.f.}$$

$$\text{between-sample mean square} = 62 \text{ with 3 d.f.}$$

If the null hypothesis is correct, these two estimates of σ_0^2 should not differ significantly. If it is incorrect, the between-sample estimate of σ_0^2 will be greater than the within-sample estimate because of between-sample variation. To test whether it is significantly greater, a one-tailed *F*-test is used (see Section 3.6):

$$F_{3,8} = 62/3 = 20.7$$

(Remember each mean *square* is used so no further squaring is necessary). From Table A.2 the critical value of *F* is 4.066 ($P = 0.05$). Since the calculated value of *F* is greater than this the null hypothesis is rejected: the sample means *do* differ significantly.

A significant result in one-way ANOVA can arise for several different reasons: for example, one mean may differ from all the others, all the means may differ from each other, the means may fall into two distinct groups, etc. A simple way of deciding the reason for a significant result is to arrange the means in increasing order and compare the difference between adjacent values with a quantity called the **least significant difference**. This is given by $s\sqrt{(2/n)} \times t_{h(n-1)}$ where s is the within-sample estimate of σ_0 and $h(n-1)$ is the number of degrees of freedom of this estimate. For the example above, the sample means arranged in increasing order of size are:

$$\bar{x}_D = 92 \quad \bar{x}_C = 97 \quad \bar{x}_A = 101 \quad \bar{x}_B = 102$$

and the least significant difference is $\sqrt{3} \times \sqrt{(2/3)} \times 2.306$ ($P = 0.05$), giving 3.26. Comparing this value with the differences between the means suggests that $\bar{x}_D$ and $\bar{x}_C$ differ significantly from each other and from $\bar{x}_A$ and $\bar{x}_B$ but that $\bar{x}_A$ and $\bar{x}_B$ do not differ significantly from each other, i.e. it is exposure to light which affects the fluorescence.

The least significant difference method described above is not entirely rigorous: it can be shown that it leads to rather too many significant differences. However it is a simple follow-up test when ANOVA has indicated that there *is* a significant difference between the means. Descriptions of other more rigorous tests are given in the references at the end of this chapter.

3.10 THE ARITHMETIC OF ANOVA CALCULATIONS

In the use of ANOVA to test for the difference between several means, σ_0^2 was estimated in two different ways. If the null hypothesis were true, σ_0^2 could also be estimated in a third way by treating the data as one large sample. This would involve summing the squares of the deviations from the overall mean:

$$\sum_i \sum_j (x_{ij} - \bar{x})^2 = 4^2 + 2^2 + 3^2 + 3^2 + 3^2 + 6^2 + 1^2 + 3^2 + 1^2 + 8^2 + 6^2 + 4^2$$

$$= 210$$

and dividing by the number of degrees of freedom, $12 - 1 = 11$.

This method of estimating σ_0^2 is not used in the analysis because the estimate depends both on the within- and between-sample variation. However, there is an exact algebraic relationship between this total variation and the sources of variation which contribute to it, which, especially in more complicated ANOVA calculations, leads to a simplification of the arithmetic involved. The relationship between the sources of variation is illustrated by Table 3.3, which summarizes the sums of squares

Table 3.3.   Summary of sums of squares and degrees of freedom

Source of variation	Sum of squares	Degrees of freedom
Between-sample	$n\sum_i(\bar{x}_i - \bar{x})^2 = 186$	$h - 1 = 3$
Within-sample	$\sum_i\sum_j(x_{ij} - \bar{x}_i)^2 = 24$	$h(n-1) = 8$
Total	$\sum_i\sum_j(x_{ij} - \bar{x})^2 = 210$	$hn - 1 = 11$

and degrees of freedom. It will be seen that the values for the total variation given in the last row of the table are the sums of the values in the first two rows for both the sum of squares and the degrees of freedom. This additive property holds for all the ANOVA calculations described in this book.

Just as in the calculation of variance, there are formulae which simplify the calculation of the individual sums of squares. These formulae are summarized in Table 3.4, which uses the notation above and also introduces the symbols:

$$\text{total number of measurements} = N = nh$$
$$\text{sum of the measurements in the } i\text{th sample} = T_i$$
$$\text{sum of all the measurements, grand total} = T$$

Table 3.4. Formulae for one-way ANOVA calculations

Source of variation	Sum of squares	Degrees of freedom
Between-samples	$\sum_i T_i^2/n - T^2/N$	$h - 1$
Within-samples	by subtraction	by subtraction
Total	$\sum_i \sum_j x_{ij}^2 - T^2/N$	$N - 1$

The use of the formulae in Table 3.4 can be illustrated by repeating the ANOVA calculations for the data in Table 3.1. The calculation of the mean squares is set out below. All the values in Table 3.1 have had 100 subtracted from them, which simplifies the arithmetic considerably. Note that this does not affect either the between- or within-sample estimates of variance because the same quantity has been subtracted from *every* value.

				T_i	T_i^2
A	2	0	1	3	9
B	1	1	4	6	36
C	−3	−5	−1	−9	81
D	−10	−8	−6	−24	576
				$T = -24$	$\sum_i T_i^2 = 702$

$$n = 3 \quad h = 4 \quad N = 12 \quad \sum_i \sum_i x_{ij}^2 = 258$$

Source of variation	Sum of squares	Degrees of freedom	Mean square
Between-sample	$702/3 - (-24)^2/12 = 186$	3	$186/3 = 62$
Within-sample	by subtraction $= 24$	8	$24/8 = 3$
Total	$258 - (-24)^2/12 = 210$	11	

Certain assumptions have been made in performing the ANOVA calculations above. The first is that the variance of the random error is not affected by the treatment used. This assumption is implicit in the pooling of the within-sample variances to calculate an overall estimate of the error variance. In doing this we are assuming what is known as the **homogeneity of variance**. In the particular example given above, where all the measurements are made in the same way, we would expect homogeneity of variance. Various methods of testing for this property are given in the references at the end of this chapter.

A second assumption is that the uncontrolled variation is random. This would not be the case if, for example, there were some uncontrolled factor, such as temperature change, which produced a trend in the results over a period of time. The effect of such uncontrolled factors can be overcome to a large extent by the techniques of randomization and blocking which are discussed in Chapter 7.

It will be seen that an important part of ANOVA is the application of the F-test. Use of this test (cf. Section 3.6) simply to compare the variances of two samples, depends on the samples being drawn from a normal population. Fortunately, however, the F-test as applied in ANOVA is not too sensitive to departures from normality of distribution.

3.11 THE CHI-SQUARED TEST

The significance tests so far described in this chapter have, in general, been concerned with testing whether the mean of several observations differs significantly from the value proposed by the null hypothesis. The data used have taken the form of observations which, apart from any rounding off, have been measured on a continuous scale. In contrast, this section is concerned with *frequency*, i.e. the number of times a given event occurs. For example, Table 2.3 gives the frequencies of the different values obtained for the nitrate ion concentration when 50 measurements were made on a sample. As discussed in Chapter 2, such measurements are usually assumed to be drawn from a population which is normally distributed: the chi-squared test can be used to test whether the observed frequencies differ significantly from those which would be expected on this null hypothesis. Since the calculation involved in this case is relatively complicated it will not be described here. (A reference to a worked example is given at the end of the chapter). The principle of the chi-squared test is more easily understood by means of the following example.

Example. The numbers of glassware breakages reported by four laboratory workers over a given period are shown below. Is there any evidence that the workers differ in their reliability?

Numbers of breakages: 24, 17, 11, 9

The null hypothesis adopted is that there is no difference in reliability. Assuming that the workers use the laboratory for an equal length of time, we would expect, from the null hypothesis, the same number of breakages by each worker. Since the total number of breakages is 61, the expected number of breakages per worker is $61/4 = 15.25$. Obviously it is not possible in practice to have a non-integral number of breakages: this number is a mathematical concept. The nearest practicable 'equal' distribution is $15, 15, 15, 16$, in some order. The question to be answered is whether the difference between the observed and expected frequencies is so large that the null hypothesis should be rejected. That there should be *some* difference between the two sets of frequencies can be most readily appreciated by considering a sequence of throws of a die: we should, for example, be most surprised if 30 throws of a die yielded exactly equal frequencies for 1, 2, 3, etc. The calculation of chi-squared, χ^2, the quantity used to test for a significant difference, is shown below.

Observed frequency, O	Expected frequency, E	$O-E$	$(O-E)^2/E$
24	15.25	8.75	5.020
17	15.25	1.75	0.201
11	15.25	−4.25	1.184
9	15.25	−6.25	2.561
		0.00	$\chi^2 = 8.966$

Note that the total of the $O-E$ column is always zero, thus providing a useful check on the calculation.

If χ^2 exceeds a certain critical value the null hypothesis is rejected. The critical value depends, as in other significance tests, on the significance level of the test and on the number of degrees of freedom. The number of degrees of freedom is, in an example of this type, one less than the number of classes used, i.e. $4-1=3$ in this case. The critical values of χ^2 for $P=0.05$ are given in Table A.5. For 3 degrees of freedom the critical value is 7.81. Since the calculated value of χ^2 is greater than this the null hypothesis is rejected at the 5% significance level: there is evidence that the workers *do* differ in their reliability.

In this calculation of χ^2 it appears that a significant result is obtained because of the high number of breakages reported by the first worker. To study this further, additional chi-squared tests can be performed. One of them tests whether the second, third and fourth workers differ significantly from each other: in this case each expected frequency is $(17+11+9)/3$. (Note that the t-test cannot be used here as we are dealing with frequencies and not continuous variates). Another tests whether the first worker differs significantly from the other three workers taken as a group. In this case there are two classes: the breakages by the first worker with an expected frequency of 15.25 and the total breakages by the other workers with expected frequency of $15.25 \times 3 = 45.75$. In such cases when there are only two classes and hence one degree of freedom, an adjustment known as **Yates's correction** should be applied. This involves replacing $O-E$ by $|O-E|-0.5$, e.g. −4.5 becomes 4. These further tests are given as an exercise at the end of this chapter.

In general the chi-squared test should only be used if the total number of observations is 50 or more and the individual expected frequencies are not less than 5. This is not a rigid rule: a reference is given at the end of this chapter which discusses this point further. Other applications of the chi-squared test are also described in this reference.

3.12 TESTING FOR NORMALITY OF DISTRIBUTION

As has been emphasized in this chapter, many statistical tests assume that the data used are drawn from a normal population. One method of testing this assumption, viz. using the chi-squared test, was mentioned in the previous section. Unfortunately, this method can only be used if there are 50 or more data points. It is common in experimental work to have only a small set of data. A simple visual way of seeing whether a such a set of data is consistent with the assumption of normality is to plot a **cumulative frequency curve** on special graph paper known as **normal probability paper**. This method is most easily explained by means of an example.

Example. Use normal probability paper to test whether the data below could have been drawn from a normal population:

109, 89, 99, 99, 107, 111, 86, 74, 115, 107, 134, 113, 110, 88, 104.

Table 3.5

Measurement	Cumulative frequency	% Cumulative frequency
74	1	6.3
86	2	12.5
88	3	18.8
89	4	25.0
99	6	37.5
104	7	43.8
107	9	56.3
109	10	62.5
110	11	68.8
111	12	75.0
113	13	81.3
115	14	87.5
134	15	93.8

Table 3.5 shows the data arranged in order of increasing size. The second column gives the cumulative frequency for each measurement, i.e. the number of measurements less than or equal to that measurement. The third column gives the percentage cumulative frequency. This is calculated by using the formula:

$$\% \text{ cumulative frequency} = 100 \times \text{cumulative frequency}/(n+1)$$

where n is the total number of measurements (there are mathematical reasons for dividing by $n+1$ rather than by n). If the data come from a normal distribution, a graph of percentage cumulative frequency against measurement results in an S-shaped curve, as shown in Fig. 3.2.

Normal probability paper has a non-linear scale for the percentage cumulative frequency axis, which will convert this S-shaped curve into a straight line. A graph plotted on such paper is shown in Fig. 3.3: the points lie approximately on a straight line, supporting the hypothesis that the data come from a normal distribution.

One method of testing for normality is to use a quantity which measures how closely the points on a normal probability plot conform to a straight line. The calculation of this quantity, the product-moment correlation coefficient, r, is described in a later chapter (Section 5.3). A reference to the use of r for testing for normality is given at the end of this chapter. This reference also gives a survey of the different tests for normality. Section 6.11 describes another method, the Kolmogorov–Smirnov method, which, among other applications, may be used to test for normality. A worked example is given in that section.

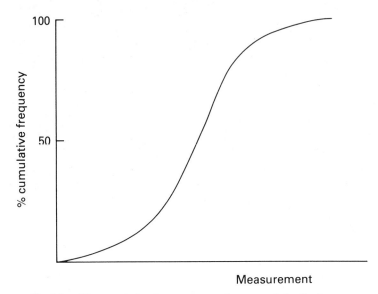

Fig. 3.2.   The cumulative frequency curve for a normal distribution.

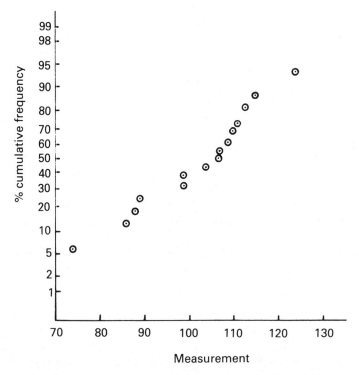

Fig. 3.3.   Normal probability plot for the example in Section 3.12.

3.13 CONCLUSIONS FROM SIGNIFICANCE TESTS

So far this chapter has concentrated on different types of significance test. This section looks more closely at the conclusions which may be drawn from a significance test. As was explained in Section 3.2, a significance test at, for example, the $P=0.05$ level involves a 5% risk that a null hypothesis will be rejected *even though it is true*. This type of error is known as a **Type 1 error**: the risk of such an error can be reduced by altering the significance level of the test to $P=0.01$ or even $P=0.001$. This, however, is not the only possible type of error: it is also possible to *retain* a null hypothesis *even when it is false*. This is called a **Type 2 error**. In order to calculate the probability of this type of error it is necessary to postulate an alternative to the null hypothesis, known as an **alternative hypothesis**.

Consider the situation where a certain chemical product is meant to contain 3% of phosphorus by weight. It is suspected that this proportion has increased and to test this the composition is analysed by a standard method with a known standard deviation of 0.03%. Suppose 4 measurements are made and a significance test is performed at the $P=0.05$ level. A one-tailed test is required, as we are interested only in an *increase*. The null hypothesis is:

null hypothesis, $\mu=3.0\%$

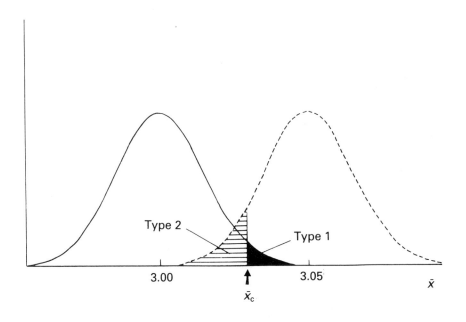

Fig. 3.4.   Type 1 and Type 2 errors.

The solid line in Fig. 3.4 shows the sampling distribution of the mean if the null hypothesis is true. This sampling distribution has mean 3.0 and standard deviation (i.e. standard error of the mean) $\sigma/\sqrt{n} = 0.03/\sqrt{4}\%$. If the sample mean lies above the indicated critical value, $\bar{x}_c$, the null hypothesis is rejected. Thus the shaded region, with area 0.05, represents the probability of a Type 1 error.

Suppose we take the alternative hypothesis:

alternative hypothesis, $\mu = 3.05\%$.

The broken line in Fig. 3.4 shows the sampling distribution of the mean if the alternative hypothesis is true. *Even if this is the case*, the null hypothesis will be retained if the sample mean lies below $\bar{x}_c$. The probability of this Type 2 error is represented by the hatched area. The diagram makes clear the interdependence of the two types of error. If, for example, the significance level is changed to $P=0.01$ in order to reduce a risk of a Type 1 error, $\bar{x}_c$ will be increased and the risk of a Type 2 error is also increased. Conversely, a decrease in the risk of a Type 2 error can only be achieved at the expense of an increase in the probability of a Type 1 error. The only way in which both errors can be reduced (for a given alternative hypothesis) is by increasing the sample size. The effect of increasing n to 9, for example, is illustrated in Fig. 3.5: the resultant decrease in the standard error of the mean produces a decrease in both types of error, for a given value of $\bar{x}_c$.

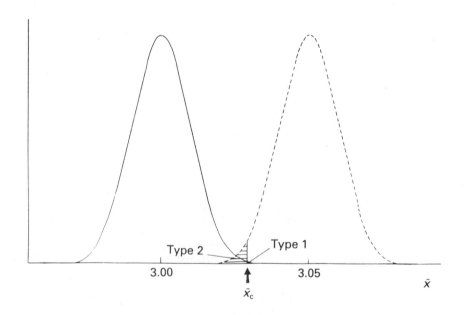

Fig. 3.5.   Type 1 and Type 2 errors for increased sample size.

The probability that a false null hypothesis is rejected is known as the **power** of a test. That is, the power of a test is (1 − the probability of a type 2 error). In the example above it is a function of the mean specified in the alternative hypothesis, and depends also on the sample size, the significance level of the test, and whether the test is one- or two-tailed. In some circumstances where two or more tests are available to test the same hypothesis, it may be useful to compare the powers of the tests in order to decide which is most appropriate.

Type 1 and Type 2 errors are also relevant when significance tests are applied sequentially. An example of this situation is the application of the t-test to the difference between two means, after first using the F-test to decide whether or not the sample variances can be pooled (see Sections 3.3 and 3.6). Both Type 1 and Type 2 errors can arise from the initial F-test, and the occurrence of either type will mean that the stated levels of significance for the subsequent t-test are incorrect, because the incorrect form of the t-test has been applied.

This example emphasizes the general conclusion that significance tests do not give clear-cut answers: rather they aid the interpretation of experimental data by giving the probabilities that certain conclusions are valid.

BIBLIOGRAPHY

V. Barnett and T. Lewis, *Outliers in Statistical Data*, 2nd Ed., Wiley, New York, 1984. A very comprehensive treatment of the philosophy of outlier rejection and the tests used.

G. E. P. Box, W. G. Hunter and J. S. Hunter, *Statistics for Experimentalists*, Wiley, New York, 1978. Gives further details of testing for significant differences between means as a follow-up to ANOVA.

R. C. Campbell, *Statistics for Biologists*, Cambridge University Press, Cambridge, 1974. Gives tests for homogeneity of variance and normality.

O. L. Davies and P. L. Goldsmith, *Statistical Methods in Research and Production,* Longmans, London, 1972. Gives more detail about Type 1 and Type 2 errors and other applications of the chi-squared test.

J. J. Filliben, *Technometrics,* 1975, **17**, 111. Describes the use of r in testing for normality, and surveys other tests for normality.

B. R. Kowalski (ed.), *Chemometrics: Theory and Application,* American Chemical Society, Washington, 1977. Chapter 11 describes tests for normality and discusses the effect of non-normality on parametric tests.

J. C. Miller, *Statistics for Advanced Level,* 2nd Ed., Cambridge University Press, Cambridge, 1989. Gives an example of the chi-squared test for normality.

R. R. Sokal and F. J. Rohlf, *Biometry,* Freeman, 1969. Gives details of tests for homogeneity of variance.

EXERCISES

1. By using a normal probability plot, test whether the following set of data could have been drawn from a normal population:

11.68, 11.12, 8.92, 8.82, 10.31, 11.88, 9.84, 11.69, 9.53, 10.30, 9.17, 10.04, 10.65, 10.91, 10.32, 8.71, 9.83, 8.90, 10.40

2. The data below are taken from the example in Section 3.3 concerning the concentration of thiol in the blood lysate. Verify that 2.07 is not an outlier.

1.84   1.92   1.94   1.92   1.85   1.91   2.07

3. The following data give the recovery of bromide from spiked samples of vegetable matter, measured by using a gas–liquid chromatographic method. The same amount of bromide was added to each specimen.

Tomato: 777 790 759 790 770 758 764 μg/g
Cucumber: 782 773 778 765 789 797 782 μg/g

(J. A. Roughan, P. A. Roughan and J. P. G. Wilkins, *Analyst*, 1983, **108**, 742)

(a) Test whether the recoveries from the two vegetables have variances which differ significantly.
(b) Test whether the mean recovery rates differ significantly.

4. The following results show the percentage of the total available interstitial water recovered by centrifuging samples taken at different depths in sandstone.

Depth of sample (m)	Water recovered, %
7	33.3 33.3 35.7 38.1 31.0 33.3
8	43.6 45.2 47.7 45.4 43.8 46.5
16	73.2 68.7 73.6 70.9 72.5 74.5
23	72.5 70.4 65.2 66.7 77.6 69.8

(K. G. Wheatstone and D. Gelsthorpe, *Analyst*, 1982, **107**, 731)

Show that the percentage of water recovered differs significantly at different depths. Use the least significant difference method described in Section 3.9 to find the causes of this significant result.

5. In reading a burette to 0.01 ml the final figure has to be estimated. The following frequency table gives the final figures of 50 such readings. Carry out an appropriate significance test to determine whether some digits are preferred to others.

Digit	0	1	2	3	4	5	6	7	8	9
Frequency	1	6	4	5	3	11	2	8	3	7

6. In order to evaluate a spectrophototmetric method for the determination of titanium, the method was applied to alloy samples containing different certified amounts of titanium. The results (% Ti) are shown below.

Sample	Certified value	Mean	Standard deviation
1	0.496	0.482	0.0257
2	0.995	1.009	0.0248
3	1.493	1.505	0.0287
4	1.990	2.002	0.0212

For each alloy 8 replicate determinations were made.

(Qiu Xing-chu and Zhu Ying-quen, *Analyst*, 1983, **108**, 641)

For each alloy, test whether the mean value differs significantly from the certified value.

7. The table below shows further results from the paper cited in Section 3.3 concerning the extraction and determination of tin in foodstuffs (*Analyst*, 1983, **108**, 109). The results give the levels of tin recovered from the same product after boiling for different times in an open vessel.

Boiling time (min)	Tin found (mg/kg)
30	57, 57, 55, 56, 56, 55, 56, 55
75	51, 60, 48, 32, 46, 58, 56, 51

Test whether
(a) the variability of the results is greater for the longer boiling time
(b) the means differ significantly.

8.   The data given in the example in Section 3.12 for the number of breakages by four different workers are reproduced below:

$$24, \ 17, \ 11, \ 9$$

Test whether
(a) the number of breakages by the first worker differs significantly from those of the other three workers;
(b) the second, third and fourth workers differ significantly from each other in carefulness.

9.   The following results were obtained in a comparison between a new method and the official method of determining phenylbutazone. (The figures refer to the percentage recoveries).

	New method	Official method
Mean	99.35	99.53
Variance	0.185	0.152
Sample size	3	3

(M. M. Amer, A. H. Taha, B. A. El-Zeany and O. A. El-Sawy, *Analyst*, 1982, **107**, 908)

Test whether the mean results obtained by the two methods differ significantly.

10.   The table below gives the amount (in mg/ml) of ephedrine hydrochloride found in pharmaceutical preparations of Ephedrine Elixir B.P. by two different methods: derivative ultraviolet spectroscopy and an official assay method (the nominal amount in each sample was 3 mg/ml).

Sample No.	Derivative method	Official method
1	2.964	2.913
2	3.030	3.000
3	2.994	3.024

(A. G. Davidson and H. Elsheikh, *Analyst*, 1982, **107**, 879)

Test whether the results obtained by the two different methods differ significantly.

11.   The following figures refer to the concentration of albumin, in g/l., in the blood sera of 16 healthy adults:

$$37, \ 39, \ 37, \ 42, \ 39, \ 45, \ 42, \ 39, \ 44, \ 40, \ 39, \ 45, \ 47, \ 47, \ 43, \ 41$$

(J. W. Foote and H. T. Delves, *Analyst*, 1983, **108**, 492)

The first 8 figures are for men and the second 8 for women. Test whether the mean concentrations for men and women differ significantly.

12.   A new flame atomic-absorption spectroscopic method of determining antimony in the atmosphere was compared with the recommended colorimetric method. For samples from an urban atmosphere the following results were obtained:

	Antimony found (mg/m³)	
Sample no.	New method	Standard method
1	22.2	25.0
2	19.2	19.5
3	15.7	16.6
4	20.4	21.3
5	19.6	20.7
6	15.7	16.8

(J. R. Castillo, J. Lanaja, M. C. Marinez and J. Aznárez, *Analyst*, 1982, **107**, 1488)

Do the results obtained by the two methods differ significantly?

4

Quality control and sampling

4.1 INTRODUCTION

The statistical tests described in the previous chapter were applied to situations that were rather simpler than those actually found in analytical laboratories. Thus we assumed that no difficulty or error was involved in securing the chemical samples supplied for analysis. In practice, sampling causes problems in quite straightforward analyses. The analysis for boron in plant samples (cf. Section 3.3) might in practice be complicated if the boron levels varied in different parts of a single plant, or from one plant to another. Two other problems must also be addressed. These are the variations that occur when the same method is applied to similar sample(s) *in different laboratories*, and the application of statistics to repeated measurements on samples having characteristics which may *vary with time*, such as the successive items on a production line. In these studies conventional statistical methods (e.g. tests for outliers, *t*-tests, ANOVA etc.) are applied to situations that are very important in the development and application of analytical methods.

4.2 SAMPLING

This topic is of fundamental importance because, unless the sampling stage of an analysis is given careful consideration, the statistical methods described in this book may not be valid for application to the results. An analyst has to deal with samples since in most cases it is impracticable or impossible to analyse all of the object under consideration. For example it is not practicable to analyse a tanker full of milk for the fat content, and impossible to analyse all the water in a river for a particular pollutant. Furthermore, many analytical techniques are destructive and so cannot be applied to the whole of an object of value.

To illustrate some aspects of sampling let us consider the situation in which we have a large batch of tablets and wish to obtain an estimate for the mean weight of a tablet. Rather than weigh all the tablets, we take a few of them (say ten) and weigh each one. In this example the batch of tablets forms the population and the ten weighed tablets form a sample from this population (see Section 2.2). If the sample is to be used to deduce the properties of the population, it must be what is known statistically as a **random sample**. This is a sample taken in such a way that all the members of the population have an equal chance of being included. Only then will equations such as (2.9), which gives the confidence limits of the mean, be valid. It must be appreciated that the term 'random' has, in the statistical sense, a different meaning from 'haphazard'. Although in practice an analyst might spread the tablets on his desk and attempt to pick a sample of ten in a haphazard fashion, such a method could conceal an unconscious bias. The best way to obtain a random sample is by the use of a random number table. Each member of the population is allocated a number in such a way that all the numbers have an equal number of digits e.g. 001, 002, 003 etc. Random numbers are then read off from a random number table (see Table A.6), starting at an arbitrary point to give, for example, 964, 173 etc. and the corresponding members of the population form the sample. An alternative (and much simpler) method which is sometimes used is to select the population members at regular intervals, for example to take every hundredth tablet off a production line. This method is not entirely satisfactory since there might be a coinciding periodicity in the weight of the tablets. The importance of the randomness of the sample is evident. Again, if the last few tablets were taken and there had been a gradual decrease in weight during the production of the batch, then this sample would give an entirely erroneous value for the mean weight of the whole batch.

In the example above the population is made up of obvious discrete members which are nominally the same, viz. the tablets. Sampling from materials for which this is not true, such as rocks, powders, gases and liquids is called **bulk sampling**. If a bulk material were perfectly homogeneous then only a small portion or **test increment** would be needed to determine the properties of the bulk. In practice bulk materials are non-homogeneous for a variety of reasons. Materials such as ores and sediments consist of macroscopic particles with different compositions and these may not be uniformly distributed in the bulk. Fluids may be non-homogeneous on a molecular scale owing to concentration gradients. Such inhomogeneity can only be detected by taking a sample of test increments from different parts of the bulk. If possible this should be done randomly by considering the bulk as a collection of cells of equal size and selecting a sample of cells by using random numbers as described above.

From the random sample, the mean, $\bar{x}$, and variance, s^2, can be calculated. There are two contributions to s^2: the **sampling variance**, σ_1^2, due to differences between the members of the sample, e.g. the tablets having different weights, and the **measurement variance**, σ_0^2, e.g. random errors in weighing each tablet. The next section describes how these two contributions can be separated and estimated by using ANOVA. For bulk materials the sampling variance is dependent on the size of the test increment relative to the scale of the inhomogeneities. As the test increment size increases, the inhomogeneities tend to be averaged out and so the sampling variance decreases.

4.3   SEPARATION AND ESTIMATION OF VARIANCES BY USING ANOVA

Section 3.8 described the use of one-way ANOVA to test for differences between means when there was a possible variation due to a fixed-effect factor. In this section we consider the situation where there is a random-effect factor, viz. sampling variation; one-way ANOVA is then used not to test whether several sample means differ significantly, but to separate and estimate the different sources of variation. Table 4.1 shows the results of the purity testing of a barrelful of sodium chloride. Five

Table 4.1.   Purity testing of sodium chloride

Sample	Purity (%)	Mean
A	98.8, 98.7, 98.9, 98.8	98.8
B	99.3, 98.7, 98.8, 99.2	99.0
C	98.3, 98.5, 98.8, 98.8	98.6
D	98.0, 97.7, 97.4, 97.3	97.6
E	99.3, 99.4, 99.9, 99.4	99.5

sample increments, A–E, were taken from different parts of the barrel chosen at random, and four replicate analyses were performed on each sample. As explained above, there are two possible sources of variation: that due to the random error in the measurement of purity, given by the measurement variance, σ_0^2, and that due to real variations in the sodium chloride purity at different points in the barrel, given by the sampling variance, σ_1^2. Since the within-sample mean square does not depend on the sample mean (see Section 3.9) it can be used to give an estimate of σ_0^2. The between-sample mean square *cannot* be used to estimate σ_1^2 directly, because the variation between sample means is caused both by the random error in measurement and by possible variation in the purity. It can be shown that the between-sample mean square gives an estimate of $\sigma_0^2 + n\sigma_1^2$ (where, as in Section 3.9, n is the number of replicate measurements). However, before an estimate of σ_1^2 is made, a test should be carried out to see whether it differs significantly from 0. This is done by comparing the within- and between-sample mean squares: if they do not differ significantly then $\sigma_1^2 = 0$ and both mean squares estimate σ_0^2.

The calculation of the mean squares by using the formulae in Table 3.4 is set out below. All the values in Table 4.1 have had 98.5 subtracted from them to simplify the arithmetic.

					T_i	T_i^2
Sample A	0.3	0.2	0.4	0.3	1.2	1.44
Sample B	0.8	0.2	0.3	0.7	2.0	4.00
Sample C	−0.2	0.0	0.3	0.3	0.4	0.16
Sample D	−0.5	−0.8	−1.1	−1.2	−3.6	12.96
Sample E	0.8	0.9	1.4	0.9	4.0	16.00
					$T = 4.0$	$\sum_i T_i^2 = 34.56$

$n = 4$ $h = 5$ $N = 20$ $\sum_i \sum_j x_{ij}^2 = 9.62$

Source of variation	Sum of squares	Degrees of freedom	Mean squares
Between-sample	$34.56/4 - 4.0^2/20 = 7.84$	4	$7.84/4 = 1.96$
Within-sample	by subtraction $= 0.98$	15	$0.98/15 = 0.0653$
Total	$9.62 - 4.0^2/20 = 8.82$	19	

Since the between-sample mean square is greater than the within-sample mean square, σ_1^2 *may* differ significantly from 0. Using the F-test to compare the two mean squares gives:

$$F_{4,15} = 1.96/0.0653 = 30$$

From Table A.2 the critical value of F is 3.056 ($P = 0.05$). The calculated value of F exceeds this, so σ_1^2 differs significantly from 0. The within-sample mean square gives 0.0653 as an estimate of σ_0^2. Since the between-sample mean square estimates $\sigma_0^2 + n\sigma_1^2$ we have:

$$
\begin{aligned}
\text{estimate of } \sigma_1^2 &= (\text{between-sample mean square} - \text{within-sample mean square})/n \\
&= (1.96 - 0.0653)/4 \\
&= 0.47
\end{aligned}
$$

4.4 SAMPLING STRATEGY

If one analysis is made on each of the h sample increments (example above, Section 4.3) then the confidence limits of the mean are given by Eq. (2.9):

$$\mu = \bar{x} \pm t(s/\sqrt{h})$$

where $\bar{x}$ is the mean of the measurements and s^2 is the variance of the measurements; s^2 is an estimate of the total variance, σ^2, which is the sum of the measurement and sampling variances, i.e. $\sigma_0^2 + \sigma_1^2$ (see Sections 1.6 and 2.7), and σ^2/h (which is estimated by s^2/h) is the variance of the mean, $\bar{x}$. If the value for each sample increment is the mean of n replicate measurements, then the variance of the mean becomes $(\sigma_0^2/n + \sigma_1^2)/h = \sigma_0^2/nh + \sigma_1^2/h$. Obviously, for maximum precision, we require the variance of the mean to be as small as possible. The term due to the measurement variance can be reduced either by using a more precise method of analysis or by increasing n, the number of replicate measurements. However, there is no point in striving to make the measurement variance much less than say a tenth of the sampling variance, since any further reduction will not greatly improve the total variance (which is the sum of the two variances). Rather it is preferable to take a larger number of sample increments, since the confidence interval decreases with increasing h. If a preliminary sample has been used to estimate s, then the sample size required to achieve a given size of confidence interval can be calculated approximately (see Chapter 2, Exercise 4).

A possible sampling strategy with bulk material is to take h sample increments and blend them before making n replicate measurements. The variance of the mean of these replicate measurements is $\sigma_0^2/n + \sigma_1^2/h$. This total variance should be compared with that when each sample increment is analysed n times and the increment means are averaged, the variance then being $\sigma_0^2/nh + \sigma_1^2/h$ (see above). Obviously the latter variance is the smaller, resulting in greater precision of the mean, but more measurements (nh against h) are required. Knowledge of the values of σ_0^2 and σ_1^2 from previous experience, and the costs of sampling and analysis, can be used to calculate the cost of relative sampling strategies. In general the most economical scheme to give the required degree of precision will be used.

For bulk materials the sampling variance depends on the size of the sample increment relative to the scale of the inhomogeneities and decreases with increasing sample increment size. In some experiments it may be necessary to set an upper limit on the sampling variance so that changes in the mean can be detected. Preliminary measurements can be made to decide the minimum sample increment size required to give an acceptable level of sampling variance.

4.5   COLLABORATIVE TRIALS: INTRODUCTION

Collaborative trials seek to examine the following question: "If the same sample (or a set of identical aliquots of a single sample) is analysed by the same method in different laboratories, are the results obtained the same within the limits of random error?" All too often it is assumed that, with the use of apparently identical instruments in different laboratories, very similar results will indeed be obtained when experienced scientists apply a particular method to identical samples. Unhappily, extensive practical experience has shown repeatedly that this expectation will not be fulfilled, and that very divergent results may be obtained even when every experimental precaution seems to be taken. For example, in one study the level of polyunsaturated fatty acids in a sample of palm oil reported by 16 different laboratories varied from 5.5% to 15.0%! The large variation was all the more noteworthy in that this was not a trace analysis, and that the method used was a relatively straightforward one involving absorption spectrometry to determine the fatty acids, after the original sample had been saponified and digested with an enzyme. Again, determination of the percentage of aluminium in a sample of limestone in ten laboratories produced values ranging from 1.11% to 1.9%.

These divergent results must be taken extremely seriously. They imply that a sample of (for example) a foodstuff that apparently passes a test for safety in one laboratory might not pass if examined in another laboratory. Blood serum from an individual might seem to be quite normal in one laboratory but give indications of a pathological state in another. Such possibilities are clearly unacceptable: analytical methods that are to be used in many laboratories must be carefully chosen in order to minimize interlaboratory variation and the consequent problems of interpreting their results. Only methods which give small interlaboratory variations will generally be adopted for use in standard and reputable analyses. Much attention has thus been given to designing collaborative trials so that, with minimum effort, they yield the maximum amount of information on the sources and magnitudes of interlaboratory errors.

4.6   TWO-SAMPLE PLOTS

The alarmingly wide scatter of results obtained in different laboratories during collaborative trials might in principle be explained by the *random errors* of the measurements, or by the presence of different *systematic errors* in different laboratories. In practice the evidence is overwhelming that the variations are caused by systematic errors. This result is best demonstrated by the use of a two-sample (or *X–Y*) chart, devised by W. J. Youden. The principle involved is that each laboratory collaborating in the trial should be sent *two similar* samples (*X* and *Y*), and be asked to perform one determination on each. The results are plotted as shown in Fig. 4.1,

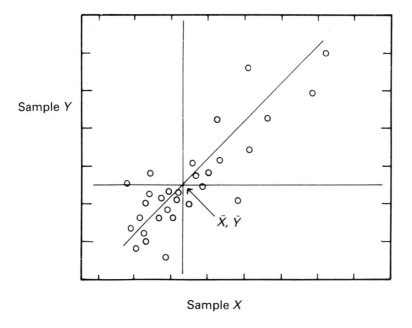

Fig. 4.1.   A Youden two-sample plot.

each point representing a pair of results from a single laboratory. The mean values of *X* and *Y* are also determined and vertical and horizontal lines are drawn through the point $(\bar{X}, \bar{Y})$, thereby dividing the chart into four quadrants.

If the results from different laboratories vary entirely because of random error, the *X* and *Y* determinations in each case may give results which are both too high, both too low, *X* high and *Y* low, or *X* low and *Y* high. These four outcomes would be equally likely, giving approximately equal numbers of points in the four quadrants of the chart. If, however, systematic errors are the main cause of the variation, we would expect that a laboratory obtaining a high value for X would also tend to obtain a high value for Y. This would lead to a predominance of points in the top right and lower left quadrants of the chart, and this is indeed the result obtained in virtually all collaborative trials. It is apparent that, in the hypothetical absence of random errors, all the points would lie on a 45° diagonal of the chart. In practice, since random errors are always present to some extent, the points fall within an ellipse which has this diagonal as its major axis. The length of the perpendicular from an individual point to the diagonal gives a measure of the random error, and the perpendicular intersects the diagonal at a point at a distance from $(\overline{X}, \overline{Y})$ which is related to the systematic error of that laboratory. It is evident that this two-sample approach to collaborative trials can give much useful information. Youden has pointed out other advantages of this approach, and we shall consider it further in Section 4.8.

4.7 PREPARING A COLLABORATIVE TRIAL

The purpose of a collaborative trial is clear — to evaluate the variation in the results of a particular analytical method (often a new or newly-modified one) when it is performed in a number of laboratories. The results might be used to decide whether the method itself is suitable for general use, and perhaps also to identify those laboratories which might be entrusted with important analytical work, e.g. with public health or other statutory implications. It thus seems that it is easy to set up such a trial. A single 'referee' laboratory (often in practice a government or other official laboratory) should send one or more samples to a number of other laboratories, along with detailed instructions on the analytical methods to be used. The collaborating laboratories will analyse the sample(s) according to these instructions, and return their results to the referee centre, where they will be evaluated by standard statistical methods. In practice, a good deal of work is involved in setting up and executing such a trial, and many important decisions must be made before any samples are dispatched.

A crucial preliminary experiment is the 'ruggedness' test. As we saw in Chapter 1, even very simple analytical experiments involve several individual steps and perhaps the use of a number of reagents. Thus many experimental factors (e.g. temperature, solvent composition, pH, humidity, reagent purity and concentration etc.) will affect the results, and it is essential that such factors are identified and studied before any collaborative trial is undertaken. In some cases a method is found to be so sensitive to small changes in one factor that is in practice so difficult to control (e.g. very high reagent purity) that the method is rejected as impracticable before a trial takes place. In other instances the trial will continue, but the collaborators will be warned of the factors to be most carefully controlled. Although a more complete discussion of experimental design is deferred to Chapter 7, it is important to indicate here that much information can be obtained from a relatively small number of experiments. Suppose it is believed that seven experimental factors (A–G) might affect the results of an analysis. These factors must be tested at (at least) two values, called **levels**, to see whether they are really significant. Thus, if temperature is thought to affect the result, we must perform preliminary experiments at two temperatures (levels) and compare the outcomes. Similarly, if reagent purity may be important, experiments with high purity and lower purity reagent batches must be done. It might thus be thought that 2^7 preliminary experiments, covering all the possible combinations of 7 factors at two levels, will be necessary. In practice, however, just eight experiments will provide important information. The two levels of the factors are called + and − , and Table 4.2 shows how these levels are set in the eight experiments, the results

Table 4.2 — Ruggedness test for seven factors

Experiment	Factors							Result
	A	B	C	D	E	F	G	
1	+	+	+	+	+	+	+	y_1
2	+	+	−	+	−	−	−	y_2
3	+	−	+	−	+	−	−	y_3
4	+	−	−	−	−	+	+	y_4
5	−	+	+	−	−	+	−	y_5
6	−	+	−	+	+	−	+	y_6
7	−	−	+	+	−	−	+	y_7
8	−	−	−	+	+	+	−	y_8

of which are called $y_1, y_2, \ldots, y_8$. The effect of altering each of the factors from its high level to its low level is easily calculated. Thus the effect of changing B from + to − is given by $(y_1 + y_2 + y_5 + y_6)/4 - (y_3 + y_4 + y_7 + y_8)/4$. When the seven differences for factors A–G have all been calculated in this way, it is easy to identify any factors that have a worryingly large effect on the results. It may be shown that any difference which is more than twice the standard deviation of replicate measurements is significant and should be further studied. This simple set of experiments, technically known as an **incomplete factorial design**, has the disadvantage that interactions between the factors cannot be detected. This point is further discussed in Chapter 7.

Another point that has to be established before the trial begins is the number of participating laboratories. Clearly this should not be too small, otherwise the results will have little general application. Since systematic differences between laboratories will cause most of the variation in the results, it is clearly better to send a few samples to numerous laboratories than to send many samples to a few, though use of a large number of laboratories will be costly and time-consuming. In practice, many collaborative trials are satisfactorily conducted with 10–20 laboratories. If the laboratories involved are known to the referee centre, it may be right to select deliberately a range of laboratories varying in quality from excellent to weak. Alternatively, all the available laboratories can be numbered, and the participating ones selected by using a random number table (see above, Section 4.2). It is not, of course, permissible to select only top-class laboratories.

The preparation, packaging and dispatch of the samples is another important element of the trial. It is normal to have a reservoir of sample, and divide it up between the participating laboratories (some of it is naturally retained by the referee to permit any additional necessary experiments). If the sample is not naturally homogeneous, the sampling problems discussed in preceding sections may arise.

4.8 CALCULATIONS IN COLLABORATIVE TRIALS

When the experimental work of a collaborative trial is complete, and the data are reported to the referee, the statistical evaluation of the results must be performed. These calculations should reveal the extent to which the variation between partici-pating laboratories is due to systematic errors. In this section we consider the calculations involved in two very simple types of collaborative trial, and also some of the problems that may arise in practice.

Perhaps the simplest type of trial is that in which *each participating laboratory is sent a single sample and makes replicate measurements on it.* The results reported to the referee must then be analysed to separate the variation between the replicates from the variation between the laboratories. The following example shows that this can be done by a simple application of one-way ANOVA: in this case we have a random-effect factor, as described in Section 4.3.

Example. A sample of medicated animal feed containing the growth promoter olaquindox was sent to five laboratories, each of which performed duplicate analyses on it by using high-performance liquid chromatography with an ultra-violet-absorption detector. The following results were obtained:

Laboratory	Olaquindox found (mg/kg)	
1	23.0	22.2
2	23.8	22.6
3	21.0	23.8
4	26.5	27.1
5	21.4	21.4

(Data from Analytical Methods Committee, *Analyst*, 1985, **110**, 75).

Estimate the random and between-laboratory variations in this trial.

The ANOVA calculation is started by subtracting 20 from each result, giving:

Laboratory			T_i	T_i^2
1	3.0	2.2	5.2	27.04
2	3.8	2.6	6.4	40.96
3	1.0	3.8	4.8	23.04
4	6.5	7.1	13.6	184.96
5	1.4	1.4	2.8	7.84
			$T = 32.8$	$\sum_i T_i^2 = 283.84$

$$n = 2 \quad h = 5 \quad N = 10 \quad \sum_i \sum_j x_{ij}^2 = 147.06$$

The ANOVA table is then:

Source of variation	Sum of squares		Degrees of freedom	Mean square
Between-laboratory	$(283.84/2) - (32.8)^2/10$	$= 34.336$	4	8.584
Within-laboratory	by subtraction	$= 5.140$	5	1.028
Total	$147.06 - (32.8)^2/10$	$= 39.476$	9	

Since the between-laboratory mean square is greater than the within-laboratory mean square the F-test is used to see whether this difference is significant. The F value in this case is given by:

$$F_{4,5} = 8.584/1.028 = 8.350$$

The critical value of $F_{4,5}$ ($P = 0.05$, Table A.2) is 5.192, so it is concluded that the difference between the two mean squares is significant. This means that the between-laboratory systematic variation (σ_1^2) is significantly greater than 0, and can be estimated as [(between-sample mean square) − (within-sample mean square)]/n (cf. Section 4.3). The resulting value for σ_1^2 is 3.878, showing clearly that most of the variation in the results is indeed due to the systematic differences between the laboratories.

In Section 4.6 it was shown that a trial in which *each laboratory makes a single measurement on each of two similar samples* provides valuable information on random and systematic errors. This approach has the further advantages that participating laboratories are not tempted to censor one or more replicate determinations, and that more materials can be studied without large numbers of experiments.

Example.   The lead levels (in ng/g) in two similar samples (A and B) of solid milk formulations for infants were determined in nine laboratories (1–9) by graphite-furnace atomic-absorption spectrometry. The results were:

Sample					Laboratories				
	1	2	3	4	5	6	7	8	9
A	35.1	23.0	23.8	25.6	23.7	21.0	23.0	26.5	21.4
B	33.0	23.2	22.3	24.1	23.6	23.1	21.0	25.6	25.0

Estimate the overall interlaboratory variation, and its random and systematic components.

This example differs from the previous one in that there is a difference between the samples as well as the differences between laboratories. In the normal way, such a situation would be dealt with by **two-way ANOVA** (see Section 7.7). However, in this instance, there are only *two* samples, deliberately chosen to be similar in their analyte content, so there is no interest in evaluating the difference between them. The calculation can therefore be set out in a way that is numerically and conceptually simpler than a complete two-way ANOVA. In performing the calculation we note that the result obtained by each laboratory for sample A may include a systematic error. The same systematic error will presumably be included in that laboratory's result for the similar sample B. The difference D ($= A - B$) will thus have this error removed, so the spread of the D

values will provide an estimate of the random or measurement errors. Similarly, A and B can be added to give T, the spread of which gives an estimate of the overall variation in the results. The measurement variance is then estimated by:

$$\sigma_0^2 = \frac{\Sigma(D_i - \bar{D})^2}{2(n-1)} \tag{4.1}$$

and the overall variance, σ^2, due to all sources of error, is estimated by:

$$\sigma^2 = \frac{\Sigma(T_i - \bar{T})^2}{2(n-1)} \tag{4.2}$$

Notice that each of these equations includes a 2 in the denominator. This is because D and T each give estimates of errors in *two* sets of results, which are subtracted and added in D and T respectively. The results of this trial can thus be expressed in a table as follows:

	1	2	3	4	5	6	7	8	9
A	35.1	23.0	23.8	25.6	23.7	21.0	23.0	26.5	21.4
B	33.0	23.2	22.3	24.1	23.6	23.1	21.0	25.6	25.0
D	2.1	− 0.2	1.5	1.5	0.1	− 2.1	2.0	0.9	− 3.6
T	68.1	46.2	46.1	49.7	47.3	44.1	44.0	52.1	46.4

Further calculations from the third and fourth rows of data show that $\bar{D} = 0.244$, $\bar{T} = 49.33$, and hence that the estimate of σ_0^2 is $(1.383)^2$, and the estimate of σ^2 is $(5.296)^2$. Hence $F_{8,8} = (5.296/1.383)^2 = 14.67$. The critical value $(P = 0.05)$ is 3.44 (Table A.2). The experimental value greatly exceeds this, so it is concluded that σ^2 is indeed significantly greater than σ_0^2. This implies, as we have seen, that the interlaboratory variation cannot be simply explained by random measurement errors, and that systematic errors between laboratories have occurred. The component of σ^2 that is due to these systematic errors, σ_1^2, is calculated from:

$$\sigma^2 = 2\sigma_1^2 + \sigma_0^2 \tag{4.3}$$

Note again the appearance of the 2 in the equation, because two sample materials are studied. Here it is a simple matter to calculate that the estimate of σ_1^2 is $(3.615)^2$.

Even very simple collaborative trials of such types are not without their problems. Sometimes it turns out that a laboratory cannot make measurements on both or all the samples sent to it, perhaps because a sample has been lost or damaged in transit, or because of gross errors in the laboratory itself. Missing data present problems of interpretation which are dealt with fully in more advanced texts.

Another frequent problem in collaborative trials is that of outliers — one or more laboratories may produce results that seem to be out of line with the rest. In the second example above, it will be apparent that both results obtained by laboratory 1 are considerably higher than the other data, and application of the Q-test (Section 3.7) shows that these results could indeed be rejected at the probability level $P = 0.05$. Recalculation of the σ_0^2 and σ^2 values for the remaining laboratories would show that, although the latter is still larger, the F-test indicates that the difference is no longer significant ($P = 0.05$). That is, if laboratory 1 is eliminated from the trial, all the remaining variation might be due to random measurement errors alone. In practice, the referee would probably seek to establish whether any particular reason could be found for this laboratory's results. Unless such an explanation was available, he would be reluctant to reject the data completely.

Many collaborative trials are far more complex than the simple types so far described. Typically, several samples might be sent to each of 10–20 laboratories, each of which would make two or more replicate measurements on each sample. Statistical evaluation of the results of such trials must take into account between-sample variations as well as random measurement errors and between-laboratory variations. Such calculations are further examined in Chapter 7.

4.9 CONTROL CHARTS

A situation frequently arises where a manufactured product is monitored over a period of time, perhaps to see whether individual product items contain on average the correct quantity of a given substance, and that there is not too much variation. One way of doing this is by taking small samples at regular intervals. Consider the particular situation in which the weight of tablets coming off a production line is monitored. Ideally the measured weight of each tablet should conform to a target value, μ_0, but in practice there will be some random variation from one tablet to the next. This variation is partly due to random error in making the measurement and partly due to real differences in weight between the tablets. Suppose that we know the size of the overall random variation, as measured by the population standard deviation σ, from previous experience. If the process is under control, i.e. the weights of tablets produced by the process *do* have have mean weight μ_0 and standard deviation, σ, then for samples of n tablets, approximately 95% of the sample means will lie within the range $\mu_0 \pm 2\sigma/\sqrt{n}$ and approximately 99.7% within the range $\mu_0 \pm 3\sigma/\sqrt{n}$ [see Chapter 2, Eqs. (2.5) and (2.7)]. A **control chart** is a graph which presents consecutive sample means in such a way that any corrective action can be taken as quickly as possible. Figure 4.2 shows a type of control chart, known as a **Shewhart chart**, for the process described above. The value for the sample mean, $\bar{x}$, is plotted against time and when the process is under control the values of $\bar{x}$ are normally distributed about μ_0. There are also two pairs of horizontal lines on the chart: the **warning lines** at $\mu_0 \pm 2\sigma/\sqrt{n}$ and the **action lines** at $\mu_0 \pm 3\sigma/\sqrt{n}$. The purpose of these lines is indicated by their names. Since the probability that a sample mean falls outside the action lines, when the process is under control, is only 0.003 (i.e. 3 in 1000), the process is usually stopped and investigated if this happens. The probability of $\bar{x}$ falling between each pair of action and warning lines is

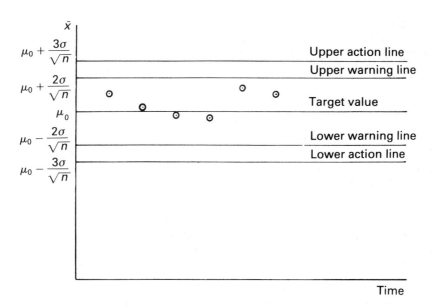

Fig. 4.2.   A typical Shewhart chart.

about 0.025 (i.e. 1 in 40): one such point would not give cause for concern but if two such points are obtained consecutively then the process should be stopped. It can be seen that a control chart displays a series of significance tests with the warning and action lines corresponding to the critical values for $P = 0.05$ and $P = 0.003$ respectively. Occasionally the process will be stopped when it is still under control (a Type 1 error — see Section 3.13) but the risk is low.

There are two reasons why points might start to fall outside the action lines. First the mean of the items produced, known as the process mean, may have drifted away from the target value; secondly there may have been an increase in variation so that the action and warning lines do not indicate the correct probabilities but are too close to the target value. Conversely a *decrease* in variation means that a change in the process mean could go undetected, again because the warning and action lines do not indicate the correct probabilities and are too far from the target value. For these reasons it is important to monitor the variation in a process as well as its mean value. This is most easily done by using the **range**, w, (i.e. the difference between the highest and lowest values) of each sample taken. The appropriate Shewhart chart can be constructed with the aid of tables which give the action and warning lines, and the target value of w for different values of n and σ (see bibliography to this chapter). The charts for mean and range are used simultaneously, appropriate action being taken if points fall outside the lines on either chart.

We have not yet considered in detail the establishment of a value for σ; this is clearly crucial in setting up control charts for both $\bar{x}$ and w; σ is sometimes called the **process capability** because it measures the inherent variation. Since it is a *population* standard deviation it must be estimated from a very large sample of, for example, tablets. This could clearly cause difficulties: we need to be sure that, when these measurements are made, there is no drift in the process mean which would give an erroneously high value for σ, but we have not yet constructed the control chart to monitor the mean! This circular problem is overcome by taking a large number of small samples over a period of time. From each sample a within-sample estimate of variance can be calculated, which is not affected by any possible variation between the samples. These estimates can then be averaged to give a value for the process capability [see Section 3.9 (i)]. In practice, the ranges of the samples are generally used, since this leads to simpler arithmetic, but the principle involved is the same.

An important consideration in using a Shewhart chart is how quickly a change in the process mean can be detected. When the process mean changes, the next point will not necessarily fall outside the action or warning lines. For example, if the process mean changes by $3\sigma/\sqrt{n}$ then there is a probability of 1/2 that the next point will fall outside the action lines; if the process mean changes by $1\sigma/\sqrt{n}$ this probability falls to 1/40. The average number of points which have to be plotted before a change in the process mean is detected is known as the **average run length (A.R.L.)**. Obviously it depends on the size of the change in the process mean compared with $\sigma/\sqrt{n}$: the greater the change the more quickly it is detected. For example, if the process mean changes by $1\sigma/\sqrt{n}$, then the A.R.L. before a sample mean falls outside the action lines is about 50. If the process is also stopped if two consecutive points fall outside the warning lines the A.R.L. is approximately halved. Since the A.R.L. depends on the change in the process mean compared with $\sigma/\sqrt{n}$, it can be reduced by increasing the sample size: the largest practicable sample size is usually dictated by considerations of cost and time: typically samples of size 4–6 are taken.

A different type of control chart, known as a **cusum chart**, utilizes all the previous sample means rather than just the last one or two as a Shewhart chart does. 'Cusum' is an abbreviation for the 'cumulative sum' i.e. the sum of the deviations of the sample means from the target value, carried forward cumulatively. An example should make its calculation clear.

Example.   Table 4.3 gives the value of the mean for a sequence of samples. The value of $\sigma/\sqrt{n}$ is known to be 2.5 and Fig. 4.3 shows the Shewhart chart for the sample means. It can be seen that although no points fall outside the warning lines, a sequence of values falls on one side of the target value.

Table 4.3 also shows the calculation of the cusum. Obviously, if the process is under control, positive and negative deviations are equally likely, so the cusum should oscillate about zero. The values of the cusum are plotted in Fig. 4.4. A good visual impact is given if the cusum chart is drawn so that the distance corresponding to $2\sigma/\sqrt{n}$ on the vertical axis is equal to the distance between successive observations on the horizontal axis.

From the cusum chart it appears that the process mean changed after the eighth sample was taken. This is one advantage of the cusum chart — it indicates at what point the process went out of control.

In order to test whether a trend in a cusum chart *does* indicate that the process

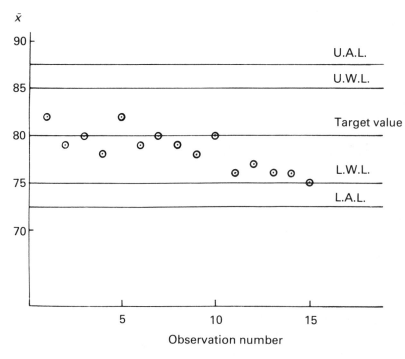

Fig. 4.3.   Shewhart chart for the data of Table 4.3 (U.A.L. stands for upper action line, and so on).

Table 4.3.   Calculation of cumulative sum: target value = 80; $\sigma/\sqrt{n}$ (known) = 2.5

Observation number	Sample mean	(Sample mean) − (Target value)	Cusum
1	82	+ 2	+ 2
2	79	− 1	+ 1
3	80	0	+ 1
4	78	− 2	− 1
5	82	+ 2	+ 1
6	79	− 1	0
7	80	0	0
8	79	− 1	− 1
9	78	− 2	− 3
10	80	0	− 3
11	76	− 4	− 7
12	77	− 3	− 10
13	76	− 4	− 14
14	76	− 4	− 18
15	75	− 5	− 23

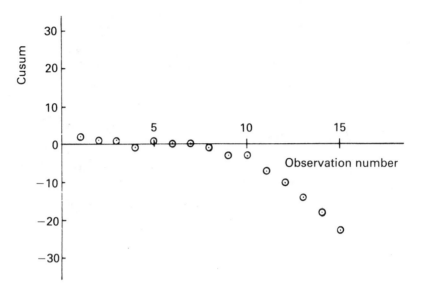

Fig. 4.4.   Cusum chart for the data of Table 4.3.

mean has changed and cannot be accounted for simply by random variation, a **V-mask** can be used. As shown in Fig. 4.5, a V-shaped mask, preferably engraved on clear plastic, is placed on the cusum chart with its axis horizontal and its apex a distance d to the right of the last observation. The semi-angle between the arms of the V is denoted by θ. The process is taken to be under control if all the values of the cusum fall between the arms of the V as shown in Fig. 4.5. Figure 4.6 illustrates a situation in which the process is out of control: two of the cusum values lie outside the upper arm of the V, which indicates that the process mean has fallen below the target value (see Figs. 4.3 and 4.4). Obviously the performance of the mask depends on the values taken for θ and d. The greater the value of d (and the smaller the value of θ) the more likely it is that a given value will fall outside the arms of the V. The values of d and θ need to be chosen so that very few false alarms occur when the process is under control, but an important change in the process mean is quickly detected. Further details of the dependence of the A.R.L. on d and θ for different changes in the process mean are given in an I.C.I. monograph cited in the bibliography to this chapter. In general the A.R.L. for a cusum chart is lower than for a Shewhart chart with the same sample size.

A cusum chart can also be used to estimate the size of the change which has occurred in the process mean when the process has gone out of control. If, for example, the process mean decreases by Δ then on average each sample mean will be Δ less than the target value. As a result the cusum will decrease on average by Δ for each point plotted. Thus the average slope of the line joining the cusum points gives a measure of the change in the process mean and hence the correction needed.

The methods described in this section can also be used for monitoring the accuracy and precision of routine analyses in the laboratory. At fixed time intervals a

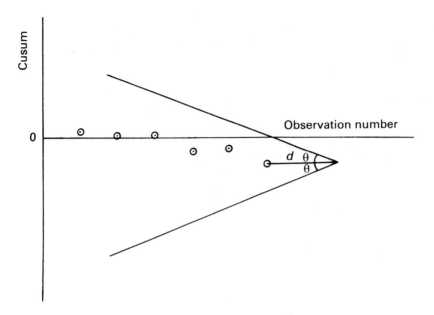

Fig. 4.5.   Use of a V-mask with the process in control.

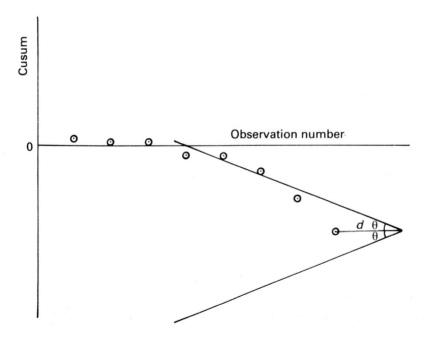

Fig. 4.6.   Use of a V-mask with the process out of control.

small number of replicate measurements are made on a standard sample and the mean and range of the replicate measurements plotted on control charts. The control chart for the mean monitors the accuracy, the target value being the known concentration of analyte in the standard. That for the range monitors the precision and the target value is the process capability. Again it is necessary to establish a value for the process capability in order to set up the control charts. In this case the process capability will almost certainly be limited by the random error involved in making the measurements rather than the error involved in preparing the standard. This can be verified by preparing a number of standards, each containing the same (nominal) analyte level, and making a number of replicate measurements on each standard. ANOVA can be used to verify that the between-standard variance is not significant compared with the random error of measurement and also to estimate the latter. Here 'replicate' means a repeated measurement made in such a way that *all* stages of the analysis are repeated, thus estimating *all* possible contributions to the random error.

BIBLIOGRAPHY

Manual on Presentation of Data and Control Chart Analysis, 6th Ed., ASTM Manual Series, Philadelphia, 1989. Gives very detailed examples of setting up Shewart charts.

D. C. Montgomery, *Introduction to Statistical Quality Control*, 2nd Ed., Wiley, New York, 1991. A detailed treatment of all aspects of quality control.

R. H. Woodward and P. L. Goldsmith, *Cumulative Sum Techniques*, I.C.I. Monograph No.3, Oliver and Boyd, Edinburgh, 1964. A detailed treatment of cusum charts.

J. Murdoch, *Control Charts*, Macmillan Press Ltd, London, 1979. Includes a detailed treatment of cusum charts.

T. F. Hartley, *Computerized Quality Control: Programs for the Analytical Laboratory*, Ellis Horwood, Chichester, 1987. Gives computer programs for use in quality control: the programs are also available in disc form.

G. Kateman and F. W. Pijpers, *Quality Control in Analytical Chemistry*, Wiley, New York, 1981. Gives a very detailed treatment of sampling.

W. J. Youden and E. H. Steiner, *Statistical Manual of the Association of Official Analytical Chemists*, Association of Official Analytical Chemists, Arlington, Virginia, 1975. Simple and well-written classic manual on collaborative trials.

G. T. Wernimont, *Use of Statistics to Develop and Evaluate Analytical Methods*, W. Spendley (ed.), Association of Official Analytical Chemists, Arlington, Virginia, 1985. Designed as an updated successor to Youden and Steiner (see above).

Analytical Methods Committee, *Analyst*, 1987, **112**, 679. This paper makes recommendations for the conduct of cooperative trials, i.e. inter-laboratory trials in which no particular analytical technique is prescribed.

EXERCISES

1. Two sampling schemes are proposed for a situation in which it is known, from past experience, that the sampling variance is 10 and the measurement variance 4 (arbitrary units).

Scheme 1: Take 5 sample increments, blend them and perform a duplicate analysis.
Scheme 2: Take 3 sample increments and perform a duplicate analysis on each.

Show that the variance of the mean is the same for both schemes.

What ratio of the cost of sampling to the cost of analysis must be exceeded for the second scheme to be the more economical?

2. The data in the table below give the concentrations of albumin measured in the blood serum of one adult. On each of four consecutive days a blood sample was taken and three replicate determinations of the serum albumin concentration were made.

Day	Albumin concentration (normalized, arbitrary units)		
1	63	61	62
2	57	56	56
3	50	46	46
4	57	54	59

Show that the mean concentrations for different days differ significantly. Estimate the variance of the day-to-day variation (i.e. the 'sampling variation').

3.   In order to estimate the measurement and sampling variances when the halofuginone concentration in chicken liver is determined, four sample increments were taken from different parts of the liver and 3 replicate measurements were made on each. The following results were obtained (mg/kg):

Sample	Replicate measurements		
A	0.25	0.22	0.23
B	0.22	0.20	0.19
C	0.19	0.21	0.20
D	0.24	0.22	0.22

Verify that the sampling variance is significantly greater than the measurement variance and estimate both variances.

Two possible sampling schemes are proposed:

Scheme 1:   Take 6 sample increments, blend them and make 4 replicate measurements.
Scheme 2:   Take 3 sample increments and make 2 replicate measurements on each.

Calculate the total variance of the mean for each scheme.

4.   In order to estimate the capability of a process, measurements were made on 6 samples of size 4 as shown in the table below (in practice at least 25 such samples would be needed). Estimate the process capability, σ. If the target value is 50, calculate the positions of the action and warning lines for a Shewhart chart for the sample mean.

Sample	Values			
1	48.8	50.8	51.3	47.9
2	48.6	50.6	49.3	49.7
3	48.2	51.0	49.3	50.3
4	54.8	54.6	50.7	53.9
5	49.6	54.2	48.3	50.5
6	54.8	54.8	52.3	52.5

5.   In a collaborative trial, two closely similar samples of oil shale (A and B) were sent to 15 laboratories, each of which performed a single inductively coupled plasma determination of the cadmium level in each sample. The following results were obtained:

Laboratory	Cd levels (ppm)	
	A	B
1	8.8	10.0
2	3.8	4.7
3	10.1	12.1
4	8.0	11.0
5	5.0	4.7
6	5.2	6.4
7	6.7	8.7
8	9.3	9.6
9	6.9	7.5
10	3.2	2.8
11	9.7	10.4
12	7.2	8.3
13	6.5	6.8
14	9.7	7.2
15	5.0	6.0

Plot the two-sample chart for these data, and comment on the principal source of error in the collaborative trial. Estimate the overall variance, the measurement variance, and the systematic error component of the variance of the results.

5

Errors in instrumental analysis; regression and correlation

5.1 INSTRUMENTAL ANALYSIS

Classical or 'wet chemistry' analysis techniques such as titrimetry and gravimetry remain in use in many laboratories and are still widely taught in Analytical Chemistry courses. They provide excellent introductions to the manipulative and other skills required in analytical work; they are ideal for high-precision analyses, expecially when small numbers of samples are involved, and they are necessary for the analysis of standard materials. However, there is no doubt that most analyses are now performed by instrumental methods. Techniques using absorption and emission spectrometry at various wavelengths, several different electrochemical methods, mass spectrometry, gas and liquid chromatography, and thermal and radiochemical methods, probably account for as much as 90% of all current analytical work, or even more. There are several reasons for this.

(1) Instrumental methods can perform analyses which are difficult or impossible by classical methods. Whereas classical methods can only rarely detect materials at sub-microgram levels, many instrumental methods are astonishingly sensitive. For example luminescence methods have detected organic molecules at the $10^{-18}M$ level. It is normally only possible to determine one analyte at a time by 'wet chemical' methods, but plasma spectometry can determine ten or more elements simultaneously (and at very low concentrations), while methods which combine high-performance liquid chromatography with a spectroscopic detection procedure can identify and determine many components of complex organic mixtures within a few minutes. Furthermore, the concentration range of a particular classical analysis method is usually limited by practical and theoretical considerations. Thus EDTA titrations can be successfully performed with reactant concentrations as low as about $10^{-4}M$, but an upper limit ($\sim 0.3M$) is set by the solubility of EDTA in water. The useful concentration range is generally 2–3 orders of magnitude (i.e. powers of ten) for classical methods. In contrast, some instrumental methods are able to determine analyte concentrations over a range of six or more orders of magnitude: this

characteristic has important implications for the statistical treatment of the results, as we shall see in the next section.

(2) For a large throughput of samples, instrumental analysis is generally quicker and often cheaper than the labour-intensive manual methods. In clinical analysis, for example, there is frequently a requirement for the same analyses to be done on scores or even hundreds of blood or blood serum/plasma samples every day. Despite the high initial cost of the equipment, such work is generally performed by using completely automatic systems. Automation has become such an important feature of analytical chemistry that the ease with which a particular technique can be automated often determines whether or not it is used at all. A typical automatic method may be able to process samples at the rate of 100 per hour or more: the equipment will take a measured volume of sample, dilute it appropriately, conduct one or more reactions on it, and determine and record the concentration of the analyte or a derivative produced in the reactions. Special problems of error estimation will evidently arise in automatic analysis; systematic errors, for example, must be identified and corrected as rapidly as possible.

(3) Modern analytical instruments are readily interfaced with computers. The present tendency is to use 16-bit or 32-bit microcomputers with floppy-disc or hard-disc storage devices. Such systems can provide a reasonably sophisticated degree of instrument control and data-handling. The computer can, for example, control the timing of sampling or injection apparatus by activating electrical switches. When the analytical results have been obtained the computer can process the data, performing such operations as the generation of derivative spectra or Fourier transforms. It can also evaluate the results statistically, the recurring theme of this book, and compare the analytical results with data stored in its memory so as to match spectral and other data. All these facilities are available from low-cost computers operating at acceptably high speeds. A further important possibility is the development of 'intelligent' instruments which, for example, perform automatic optimization processes (see Chapter 7). There is no doubt that computer facilities will soon be provided with virtually all analytical instruments, further enhancing their capabilities and efficiency.

5.2 CALIBRATION GRAPHS IN INSTRUMENTAL ANALYSIS

The ability of instrumental analysis techniques to handle a wide range of analyte concentrations means that the results are calculated, and the random errors evaluated, in a particular fashion that differs from that used when a single measurement is repeated several times. The usual procedure is as follows. The analyst takes a series of samples (normally at least three or four, and possibly several more) in which the concentration of the analyte is *known*. These calibration standards are measured in the analytical instrument under the same conditions as those subsequently used for the test (i.e. the 'unknown') samples. Once the calibration graph has been established the analyte concentration in any test sample can be obtained, as shown in Fig. 5.1, by interpolation. This general procedure raises several important statistical questions.

(1) Is the calibration graph linear? If it is a curve, what is the form of the curve?

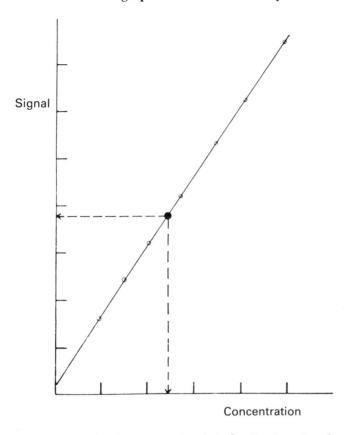

Fig. 5.1.   Calibration procedure in instrumental analysis; ○ calibration points; ● test sample.

(2) Bearing in mind that each of the points on the calibration graph is subject to errors, what is the best straight line (or curve) through these points?

(3) Assuming that the calibration plot is actually linear, what are the estimated errors and confidence limits for the slope and the intercept of the line?

(4) When the calibration plot is used for the analysis of a test sample, what are the errors and confidence limits for the determined concentration?

(5) What is the *limit of detection* of the method? That is, what is the least concentration of the analyte that can be detected with a predetermined level of confidence?

Before tackling these questions in detail, we must consider a number of aspects of plotting calibration graphs. First, it is usually essential that the calibration standards cover the whole range of concentrations required in the subsequent analyses. With the important exception of the 'method of standard additions', which is treated separately in a later section, concentrations of test samples are normally determined by interpolation and *not* by extrapolation. Secondly it is crucially important to include the value for a 'blank' sample in the calibration curve. The blank contains no deliberately added analyte, but does contain the same solvent, reagents, etc., as the other test samples, and is subjected to exactly the same sequence of analytical procedures. The instrument signal given by the blank sample will often not be zero. It is, of course, subject to errors, like all the other points on the calibration plot, and it is thus wrong in principle to subtract the blank value from the other standard values before plotting the calibration graph. Finally, it should be noted that the calibration curve is always plotted with the instrument response on the vertical (y) axis and the standard concentrations on the horizontal (x) axis. This is because many of the procedures to be described in the following sections *assume* that all the errors are in the y-values and that the standard concentrations (x-values) are error-free. This assumption is further discussed below.

The reader should note that the methods now to be described continue to generate a good deal of controversy, evidenced by many scientific meetings and original papers on the subject of calibration. Much of this controversy derives from the fact that the generally used methods rest on two assumptions. The first is that mentioned in the previous paragraph, viz. that the errors in the calibration experiment occur only in the y-values. In many routine instrumental analyses this assumption may well be justified. The standards can be made up with an error of ca. 0.1% or better (see Chapter 1), whereas the instrumental measurements themselves might have a coefficient of variation of 1–2% or worse. In recent years the advent of some high-precision automatic instruments has again put the assumption under question, however, and has led some users to make up their standard solutions by weight rather than by the less accurate combination of weight and volume.

The second assumption is that the magnitude of the errors in y is independent of the analyte concentration. Common sense indicates that this is often unlikely to be true: if, for example, the relative errors in measurement are constant, the absolute errors will increase as the analyte concentration increases. As we shall see, it is possible to modify the statistical procedure to take such factors into account; in practice, however, the modified (and rather more complex) procedures seem not to be used in many cases where they should be adopted.

5.3 THE PRODUCT-MOMENT CORRELATION COEFFICIENT

In this section we discuss the first problem listed in the previous section — is the calibration plot linear? We assume that a straight-line plot takes the algebraic form:

$$y = bx + a \tag{5.1}$$

where b is the slope of the line and a its intercept on the y-axis. The individual points on the line will be referred to as (x_1, y_1) (normally the 'blank' reading), (x_2, y_2), $(x_3, y_3) \ldots (x_i, y_i) \ldots (x_n, y_n)$, i.e. there are n points altogether. The mean of the x-values is, as usual, called $\bar{x}$, and the mean of the y-values is $\bar{y}$: the position $(\bar{x}, \bar{y})$ is then known as the 'centroid' of all the points.

In order to estimate how well the experimental points fit a straight line, we calculate the **product-moment correlation coefficient, r**. This statistic is often referred to simply as the 'correlation coefficient' because in quantitative sciences it is by far the most commonly used type of correlation coefficient. We shall, however, meet other types of correlation coefficient in Chapter 6. The value of r is given by:

$$r = \frac{\sum_i \{(x_i - \bar{x})(y_i - \bar{y})\}}{\{[\sum_i (x_i - \bar{x})^2][\sum_i (y_i - \bar{y})^2]\}^{\frac{1}{2}}} \tag{5.2}$$

Close study of this equation shows that r can take values in the range $-1 \leqslant r \leqslant +1$. As indicated in Fig. 5.2, an r value of -1 describes perfect negative correlation, i.e.

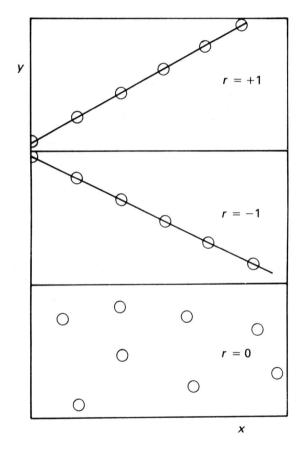

Fig. 5.2.   The product-moment correlation coefficient, r.

all the experimental points lie on a straight line of negative slope. Similarly, when $r = +1$ we have perfect positive correlation, all the points lying exactly on a straight line of positive slope. When there is no correlation between x and y the value of r is zero. In analytical practice, calibration graphs frequently give numerical r values greater than 0.99, and r values less than about 0.90 are relatively uncommon. A typical example of a calculation of r illustrates a number of important points.

Example. Standard aqueous solutions of fluorescein are examined in a fluorescence spectrometer, and yield the following fluorescence intensities (in arbitrary units):

Fluorescence intensities:	2.1	5.0	9.0	12.6	17.3	21.0	24.7
Concentration, pg/ml:	0	2	4	6	8	10	12

Determine the correlation coefficient, r.

In practice, such calculations will almost certainly be performed on a programmable calculator or microcomputer, but it is important and instructive to examine a manually calculated result. The data are presented in a table, as follows.

x_i	y_i	$x_i - \bar{x}$	$(x_i - \bar{x})^2$	$y_i - \bar{y}$	$(y_i - \bar{y})^2$	$(x_i - \bar{x})(y_i - \bar{y})$
0	2.1	-6	36	-11.0	121.00	66.0
2	5.0	-4	16	-8.1	65.61	32.4
4	9.0	-2	4	-4.1	16.81	8.2
6	12.6	0	0	-0.5	0.25	0
8	17.3	2	4	4.2	17.64	8.4
10	21.0	4	16	7.9	62.41	31.6
12	24.7	6	36	11.6	134.56	69.6
42	91.7	0	112	0	418.28	216.2

$$\bar{x} = 42/7 = 6; \bar{y} = 91.7/7 = 13.1$$

The figures below the line at the foot of the columns are in each case the sums of the figures in the table. [Note that $\sum_i (x_i - \bar{x})$ and $\sum_i (y_i - \bar{y})$ are both zero.] Using these totals in conjunction with Eq. (5.2), we have

$$r = 216.2/(112 \times 418.28)^{\frac{1}{2}}$$
$$= 216.2/216.44$$
$$= 0.9989$$

Two observations follow from this example. As is shown in Fig. 5.3, although several of the points deviate noticeably from the 'best' straight line (calculated by using the principles of the following section), the r value is very close indeed to 1. Experience shows that even quite poor-looking calibration plots give very high r values. In such cases the numerator and denominator in Eq. (5.2) are nearly equal. It is thus very important to perform the calculation with an adequate number of significant figures. In the example given, neglect of the figures after the decimal point would have given an obviously incorrect r value of exactly 1, and the use of only one place of decimals would have given the incorrect r value of 0.9991. This point is especially important when a calculator or computer is used to determine r; such devices do not always provide sufficient figures.

Although correlation coefficients are simple to calculate, they are all too easily

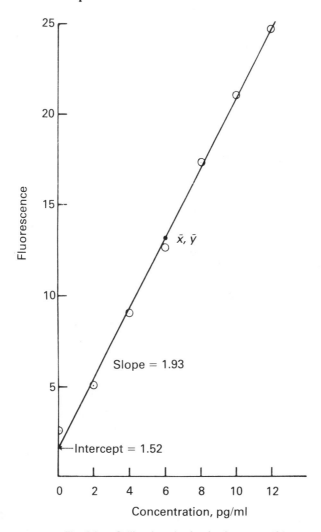

Fig. 5.3.   Calibration plot for the data on p. 106.

misinterpreted. It must always be borne in mind that the use of Eq. (5.2) will generate an r value even if the data are patently non-linear in character. Figure 5.4 shows two examples in which a calculation of r would be misleading. In Fig. 5.4a, the points of the calibration plot clearly lie on a curve; this curve is sufficiently gentle, however, to yield quite a high correlation coefficient when Eq. (5.2) is applied. The lesson of this example is that the calibration curve must *always* be physically plotted (on graph paper or a computer monitor): otherwise a straight-line relationship might wrongly be deduced from the calculation of r. Figure 5.4b is a reminder that a zero correlation coefficient does not mean that y and x are entirely unrelated; it only means that they are not *linearly* related.

As we have seen, r values obtained in instrumental analysis are normally very high, so a calculated value, together with the calibration plot itself, is often sufficient

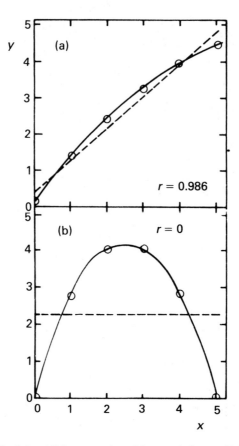

Fig. 5.4.   Misinterpretation of the correlation coefficient, r.

to assure the analyst that he has indeed obtained a useful linear relationship. In some circumstances, however, much lower r values are obtained; one such situation is further discussed in Section 5.9. In these cases it will be necessary to use a proper statistical test to see whether the correlation coefficient is indeed significant, bearing in mind the number of pairs of points used in the calculation. The simplest method of doing this is to calculate a t-value (see Chapter 3 for a fuller discussion of the t-test), using the equation

$$t = \frac{|r| \sqrt{(n-2)}}{\sqrt{(1-r^2)}}$$

(5.3)

The calculated value of t is compared with the tabulated value at the desired significance level, using a *two-tailed* t-test and $(n-2)$ degrees of freedom. The null hypothesis in this case is that there is no correlation between x and y. If the calculated value of t is greater than the tabulated value, the null hypothesis is rejected; that is, we conclude in such a case that a significant correlation does exist.

5.4 THE LINE OF REGRESSION OF y ON x

In this section we assume that there is a linear relationship between the analytical signal (y) and the concentration (x), and show how to calculate the 'best' straight line through the calibration graph points, each of which is subject to experimental error. Since we are assuming for the present that all the errors are in y (cf. Section 5.2 above), we are seeking the line that minimizes the deviations in the y-direction between the experimental points and the calculated line. Since some of these deviations (technically known as the y-residuals) will be positive and some negative, it is sensible to seek to minimize the **sum of the squares of the residuals.** This explains the frequent use of the term 'method of least squares' for the procedure. The straight line required is calculated on this principle: as a result it is found that the line must pass through the 'centroid' of the points, ($\bar{x}$, $\bar{y}$). It can be shown that

$$b = \frac{\sum_i \{(x_i - \bar{x})(y_i - \bar{y})\}}{\sum_i (x_i - \bar{x})^2} \qquad (5.4)$$

$$a = \bar{y} - b\bar{x} \qquad (5.5)$$

The line thus calculated is known as the **line of regression of** y **on** x, i.e. the line indicating how y varies when x is set to chosen values. It is very important to notice that the line of regression of x on y *is not the same line* (except in the highly improbable case where all the points lie exactly on a straight line, when $r = 1$ exactly). The line of regression of x on y (which also passes through the centroid of the points) assumes that all the errors occur in the x-direction. If we maintain rigidly the convention that the analytical signal is always plotted on the y-axis and the concentration on the x-axis, it is always the line of regression of y on x that we must use in calibration experiments.

Example. Calculate the slope and intercept of the regression line for the data given in the previous example (see Section 5.3).

In Section 5.3 we calculated that, for this calibration curve:

$$\sum_i (x_i - \bar{x})(y_i - \bar{y}) = 216.2; \quad \sum_i (x_i - \bar{x})^2 = 112; \quad \bar{x} = 6; \quad \bar{y} = 13.1.$$

Using Eqs. (5.4) and (5.5) we calculate that

$$b = 216.2/112 = 1.93$$
$$a = 13.1 - (1.93 \times 6) = 13.1 - 11.58 = 1.52$$

The equation for the regression line is $y = 1.93x + 1.52$.

It is apparent that the terms used to calculate r are also needed in the calculation of a and b: this facilitates the calculation of all these terms with a calculator or simple computer program.

The results of the slope and intercept calculations are depicted in Fig. 5.3. Again it is important to emphasize that Eqs. (5.4) and (5.5) must not be misused — they will only give useful results when prior study (calculation of r and a visual inspection of the points) has indicated that a straight line relationship is realistic for the experiment in question. Non-parametric methods (i.e. methods that make no assumptions about the nature of the error distribution) can also be used to calculate regression lines, and this topic is treated in the next chapter.

5.5   ERRORS IN THE SLOPE AND INTERCEPT OF THE REGRESSION LINE

The line of regression calculated in the previous section will in practice be used to estimate the concentrations of test samples by interpolation, and perhaps also to estimate the limit of detection of the analytical procedure. The random errors in the values for the slope and intercept are thus of importance, and the equations used to calculate them are now considered. We must first calculate the statistic $s_{y/x}$, which is given by:

$$s_{y/x} = \left\{ \frac{\sum\limits_{i} (y_i - \hat{y}_i)^2}{n - 2} \right\}^{\frac{1}{2}}$$

(5.6)

It will be seen that this equation utilizes the y-residuals, $y_i - \hat{y}_i$, where the $\hat{y}_i$ values are the points on the calculated regression line corresponding to the individual x-values, i.e. the 'fitted' y-values (Fig. 5.5). The $\hat{y}$-value for a given value of x is of course readily calculated from the regression equation. Equation (5.6) is clearly similar in form to the equation for the standard deviation of a set of repeated measurements [Eq. (2.2)]; the former differs in that deviations, $(y_i - \bar{y})$, are replaced by residuals, $(y_i - \hat{y}_i)$, and the denominator contains the term $(n - 2)$ rather than $(n - 1)$. In a linear regression calculation the number of degrees of freedom (cf. Section 2.4) is $(n - 2)$. This clearly reflects the obvious consideration that only one straight line can be drawn through two points.

Armed with a value for $s_{y/x}$, we can now calculate s_b and s_a, the standard deviations for the slope (b) and the intercept (a). These are given by:

$$s_b = \frac{s_{y/x}}{\{\sum\limits_{i} (x_i - \bar{x})^2\}^{\frac{1}{2}}}$$

(5.7)

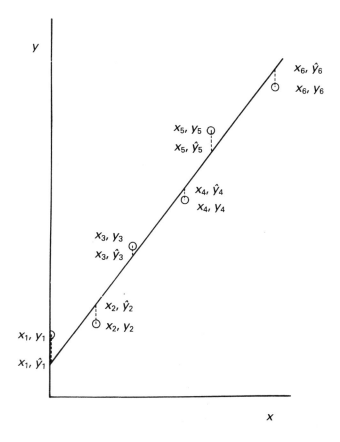

Fig. 5.5.   The y-residuals of a regression line.

$$s_a = s_{y/x} \left\{ \frac{\sum\limits_{i} x_i^2}{n \sum\limits_{i} (x_i - \bar{x})^2} \right\}^{\frac{1}{2}} \qquad (5.8)$$

The values of s_b and s_a can be used in the usual way (cf. Chapter 2) to estimate confidence limits for the slope and intercept. Thus the confidence limits for the slope are given by $b \pm t s_b$, where the t-value is taken at the desired confidence level and $(n-2)$ degrees of freedom. Similarly the confidence limits for the intercept are given by $a \pm t s_a$.

Example. Calculate the standard deviations and confidence limits of the slope and intercept of the regression line calculated in Section 5.4.

This type of calculation may not be directly accessible on a programmable calculator, although useful computer programs have been developed (see the bibliography for Chapter 1). Here we perform the calculation manually, using a tabular lay-out as before.

x_i	x_i^2	y_i	$\hat{y}_i$	$|y_i - \hat{y}_i|$	$(y_i - \hat{y}_i)^2$
0	0	2.1	1.52	0.58	0.3364
2	4	5.0	5.38	0.38	0.1444
4	16	9.0	9.24	0.24	0.0576
6	36	12.6	13.10	0.50	0.2500
8	64	17.3	16.96	0.34	0.1156
10	100	21.0	20.82	0.18	0.0324
12	144	24.7	24.68	0.02	0.0004

$$\sum_i x_i^2 = 364 \qquad\qquad\qquad \sum_i (y_i - \hat{y})^2 = 0.9368$$

From the table and using Eq. (5.6) we obtain

$$s_{y/x} = \sqrt{0.9368/5} = \sqrt{0.18736} = 0.4329.$$

From Section 5.3 we know that $\sum_i (x_i - \bar{x})^2 = 112$, and Eq. (5.7) can be used to show that

$$s_b = 0.4329/\sqrt{112} = 0.4329/10.58 = 0.0409.$$

The t-value for $(n - 2) = 5$ and the 95% confidence level is 2.57 (Table A.1). The 95% confidence limits for b are thus

$$b = 1.93 \pm 2.57 \times 0.0409 = 1.93 \pm 0.11$$

Equation (5.8) requires knowledge of $\sum_i x_i^2$, calculated as 364 from the table. We can thus write

$$s_a = 0.4329\sqrt{(364/784)} = 0.2950$$

so the confidence limits are

$$a = 1.52 \pm 2.57 \times 0.2950 = 1.52 \pm 0.76$$

In this example, the number of significant figures necessary was not large, but it is always a useful precaution to use the maximum available number of significant figures during such a calculation, rounding only at the end.

5.6 CALCULATION OF A CONCENTRATION

Once the slope and intercept of the regression line have been determined, it is very simple to calculate an x-value corresponding to any measured y-value. A more complex problem arises when it is necessary to estimate the error in a concentration calculated by using a regression line. Calculation of an x-value from a given y-value involves the use of both the slope (b) and the intercept (a) and, as we saw in the previous section, both these values are subject to error. As a result, the determination of the error in the x-value is extremely complex, and most workers use the following approximate formula:

$$s_{x_0} = \frac{s_{y/x}}{b} \left\{ 1 + \frac{1}{n} + \frac{(y_0 - \bar{y})^2}{b^2 \sum_i (x_i - \bar{x})^2} \right\}^{\frac{1}{2}}$$

(5.9)

In this equation, y_0 is the experimental value of y from which the concentration value x_0 is to be determined, s_{x_0} is the estimated standard deviation of x_0, and the other symbols have their usual meaning. In some cases an analyst may make several readings to obtain the value of y_0; if there are m such readings, then the equation for s_{x_0} becomes

$$s_{x_0} = \frac{s_{y/x}}{b} \left\{ \frac{1}{m} + \frac{1}{n} + \frac{(y_0 - \bar{y})^2}{b^2 \sum_i (x_i - \bar{x})^2} \right\}^{\frac{1}{2}}$$

(5.10)

As expected, Eq. (5.10) reduces to Eq. (5.9) if $m = 1$. As always, confidence limits can be calculated as $x_0 \pm t s_{x_0}$, $(n - 2)$ degrees of freedom. Again, a simple computer program will perform all these calculations, but most preprogrammed calculators will not be wholly adequate.

Example. Using the data from the example in Section 5.3, determine the x_0 and s_{x_0} values and x_0 confidence limits for solutions with fluorescence intensities 2.9, 13.5, and 23.0 units.

The x_0 values are easily calculated by using the regression equation determined in Section 5.4, $y = 1.93x + 1.52$. Substituting the y_0-values 2.9, 13.5 and 23.0, we obtain x_0-values of 0.72, 6.21 and 11.13 pg/ml respectively.

To obtain the s_{x_0}-values corresponding to these x_0-values we use Eq. (5.9), recalling from the preceding sections that $n = 7$, $b = 1.93$, $s_{y/x} = 0.4329$, $\bar{y} = 13.1$, and $\sum_i (x_i - \bar{x})^2 = 112$. The y_0 values 2.9, 13.5 and 23.0 then yield s_{x_0}-values of 0.26, 0.24 and 0.26 respectively. The corresponding 95% confidence limits ($t = 2.57$) are 0.72 ± 0.68, 6.21 ± 0.62, and 11.13 ± 0.68 pg/ml respectively.

This example illustrates a point of some importance. It is apparent that the confidence limits are rather smaller (i.e. better) for the result $y_0 = 13.5$ than for the other two y_0-values. Inspection of Eq. (5.9) confirms that as y_0 approaches $\bar{y}$, the third term inside the bracket approaches zero, and s_{x_0} thus approaches a minimum value. The general form of the confidence limits for a calculated concentration is shown in Fig. 5.6. In a practical analysis, therefore, a calibration experiment of this

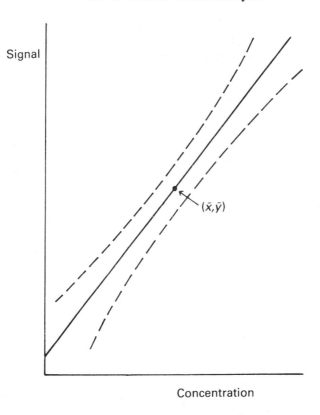

Fig. 5.6.   General form of the confidence limits for a concentation determined by using an
unweighted regression line.

type will give the most precise results when the measured instrument signal
corresponds to a point close to the centroid of the regression line.

If we wish to improve (i.e. narrow) the confidence limits in this calibration
experiment, Eqs. (5.9) and (5.10) show that at least two approaches should be
considered. We could increase n, the number of calibration points on the regression
line, and we could make more than one measurement of y_0, and use the mean value
of m such measurements in the calculation of x_0. The results of such procedures can
be assessed by considering the three terms inside the brackets in the two equations.
In the example above, the dominant term in all three calculations is the first one —
unity. It follows that in this case (and many others) an improvement in precision
might be made by measuring y_0 several times and using Eq. (5.10) rather than Eq.
(5.9). If, for example, the y_0-value of 13.5 had been calculated as the mean of four
determinations, then the s_{x_0}-value and the confidence limits would have been 0.14
and 6.21 ± 0.36 respectively, both results indicating substantially improved precision. Of course, making too many replicate measurements (assuming that sufficient
sample is available) generates much more work for only a small additional benefit:
the reader should verify that 8 measurements of y_0 would produce an s_{x_0}-value of 0.12
and confidence limits of 6.21 ± 0.30.

The effect of n, the number of calibration points, on the confidence limits of the concentration determination is more complex. This is because we also have to take into account accompanying changes in the value of t. Use of a large number of calibration samples involves the task of preparing many accurate standards for only marginally increased precision (cf. the effects of increasing m, described in the previous paragraph). On the other hand, small values of n are not permissible: in such cases, not only will $1/n$ be larger, but the number of degrees of freedom, $(n-2)$, will become very small, necessitating the use of very large t values in the calculation of the confidence limits. In many experiments, as in the example given, six or so calibration points will be adequate, the analyst gaining extra precision, if necessary, by repeated measurements of y_0.

5.7 LIMITS OF DETECTION

As we have seen, one of the principal benefits of using instrumental methods of analysis is that they are capable of detecting and determining much smaller quantities of analyte than classical analysis methods. These benefits have led to the appreciation of the importance of trace concentrations of many materials, for example in biological and environmental samples, and thus to the development of many further techniques in which low limits of detection are a major criterion of successful application. It is therefore evident that statistical methods for assessing and comparing limits of detection are of importance. In general terms, the limit of detection of an analyte may be described as that concentration which gives an instrument signal (y) *significantly different* from the 'blank' or 'background' signal. It is immediately apparent that this description gives the analyst a good deal of freedom to decide the exact definition of the limit of detection, based on a suitable interpretation of the phrase 'significantly different'. There has in practice been very little agreement between professional and statutory bodies on this point, and the whole area is as controversial as other aspects of the statistical treatment of calibration analyses. A commonly used definition in the literature of analytical chemistry is that the limit of detection is *the analyte concentration giving a signal equal to the blank signal*, y_B, *plus two standard deviations of the blank*, s_B. Recent guidelines from (particularly American) public bodies suggest that the criterion should be:

$$y - y_B = 3s_B \qquad\qquad (5.11)$$

The significance of this last definition is illustrated in more detail in Fig. 5.7. An analyst studying trace concentrations is confronted with two problems: he does not wish to claim the presence of the analyte when it is actually absent, but equally he does not wish to report that the analyte is absent when it is in fact present. The possibility of each of these errors must be minimized by a sensible definition of a limit of detection. In the figure, curve A represents the normal distribution of measured values of the blank signal. It would be possible to identify a point, $y = P$, towards the upper edge of this distribution, and claim that a signal greater than this was unlikely to be due to the blank (Fig. 5.7), whereas a signal less than P would be assumed

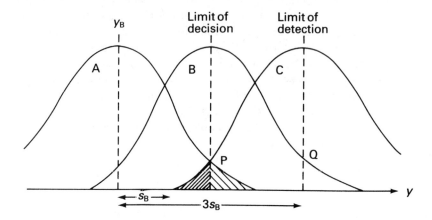

Fig. 5.7.   Definitions of (a) the limit of decision; (b) the limit of detection.

to indicate a blank sample. However, for a sample giving an average signal P, 50% of the observed signals will be less than this, since the signal will have a normal distribution (of the same shape as that for blank — see below) extending below P (curve B). The probability of concluding that this sample does not differ from the blank when in fact it *does*, is therefore 50%. Point P, which has been called the limit of decision, is thus unsatisfactory as a limit of detection, since it solves the first of the problems mentioned above, but not the second. A more suitable point is at $y = Q$ (Fig. 5.7), such that Q is twice as far as P from y_B. It may be shown that if y_B–Q is 3.28 times the standard deviation of the blank, s_B, then the probability of each of the two kinds of error occurring (indicated by the shaded areas in Fig. 5.7) is only 5%. If, as suggested by Eq. (5.11), the distance y_B–Q is only $3s_B$, the probability of each error is about 7%; many analysts would consider that this is a reasonable definition of a limit of detection.

It must be re-emphasized that this definition of a limit of detection is quite arbitrary, and it is entirely open to an analyst to provide an alternative definition for a particular purpose. For example, there may be occasions when an analyst is anxious to avoid at all costs the possibility of reporting the absence of the analyte when it is in fact present, but is relatively unworried about the opposite error. It is clear that, whenever a limit of detection is cited in a paper or report, the definition used to obtain it must also be provided. Some attempts have been made to define a further limit, the 'limit of quantitation' (or 'limit of determination'), which is regarded as the lower limit for precise quantitative measurements, as opposed to qualitative detection. A value of $y_B + 10s_B$ has been suggested for this limit, but it has not been very widely used in practice.

We must now discuss how the terms y_B and s_B are obtained in practice when a conventional regression line is used for calibration as described in the preceding sections. A fundamental assumption of the unweighted least-squares method that we have studied thus far in this chapter is that each point on the plot (including the point representing the blank or background) has a normally distributed variation (in the y-direction only) with a standard deviation estimated by $s_{y/x}$ [Eq. (5.6)]. This is the justification for drawing the normal distribution curves with the same width in Fig. 5.7. It is therefore appropriate to use $s_{y/x}$ in place of s_B in the estimation of the limit of detection. It is, of course, possible to perform the blank experiment several times and obtain an independent value for s_B, but this is a time-consuming procedure, and the use of $s_{y/x}$ is quite suitable in practice. The value of a, the calculated intercept, can be used as an estimate of y_B, the blank signal itself; it should be a more accurate estimate of y_B than the single measured blank value, y_1.

Example. Estimate the limit of detection for the fluorescein determination studied in the previous sections.

We use Eq. (5.11) with the values of y_B ($= a$) and s_B ($= s_{y/x}$) previously calculated. The value of y at the limit of detection is found to be $1.52 + 3 \times 0.4329$, i.e. 2.82. Use of the regression equation then yields a detection limit of 0.67 pg/ml. Figure 5.8 summarizes all the calculations performed on the fluorescein determination data.

It is very important to avoid confusing the limit of detection of a technique with its **sensitivity.** This very common source of confusion probably arises because there is no generally-accepted English word synonymous with 'having a low limit of detection'. The word 'sensitive' is generally used for this purpose, thus giving rise to much ambiguity. The sensitivity of a technique is correctly defined as the *slope* of the calibration graph and, provided the plot is linear, can be measured at any point on it. In contrast, the limit of detection of a method is calculated with the aid of the section of the plot close to the origin, and utilizes both the slope and the intercept.

5.8   THE METHOD OF STANDARD ADDITIONS

Suppose that an analyst wishes to determine the concentration of silver in samples of photographic waste by atomic-absorption spectrometry. Using the methods of the previous sections, he could calibrate the spectrometer with some aqueous solutions of a pure silver salt and use the resulting calibration graph in the determination of the silver in the test samples. This method is only valid, however, if a pure aqueous solution of silver gives the same absorption signal as a photographic waste sample containing the same concentration of silver. In other words, in using pure solutions to establish the calibration graph it is assumed that there are no 'matrix effects', i.e. reduction or enhancement of the silver absorbance signal by other components. In many areas of analysis, such as assumption is frequently invalid. Matrix effects occur even with methods such as plasma spectrometry that have a reputation for being relatively free from interferences.

The first possible solution to this problem is to take a sample of photographic waste that is similar to the test sample but free from silver, and add known amounts of a silver salt to it to make up the standard solutions. The calibration graph will then

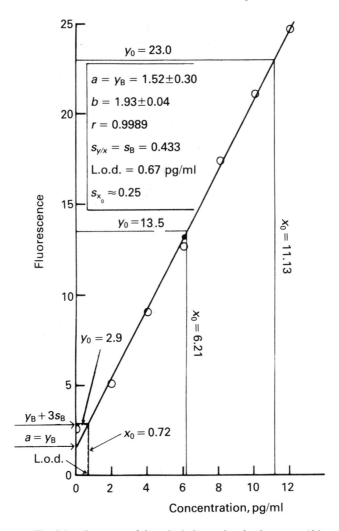

Fig. 5.8.   Summary of the calculations using the data on p. 106.

be set up by using an apparently suitable matrix. In many cases, however, this approach is impracticable. It will not eliminate matrix effects that differ in magnitude from one sample to another, and it may not be possible to obtain a sample of the matrix that contains no analyte — for example, a silver-free sample of photographic waste is unlikely to occur! It follows that all the analytical measurements, including the establishment of the calibration graph, must in some way be performed with the sample itself. This is achieved in practice by using the **method of standard additions.** The method is widely practised in atomic absorption and emission spectrometry and has also found application in electrochemical analysis and other areas. Equal volumes of the sample solution are taken, all but one are separately 'spiked' with known and different amounts of the analyte, and *all* are then diluted to the same volume. The instrument signals are then determined for all these solutions and the results plotted as shown in Fig. 5.9. As usual, the signal is plotted on the y-axis; in this case the x-axis is graduated in terms of the amounts of analyte *added* (either as an absolute weight or as a concentration). The (unweighted) regression line is calculated in the normal way, but space is provided for it to be extrapolated back to the point on the x-axis at which $y = 0$. It is clear that this negative intercept on the x-axis corresponds to the amount of the analyte in the test sample. Inspection of the figure shows that this value is given by a/b, the ratio of the intercept and the slope of the regression line. Since both a and b are subject to error (Section 5.5) the calculated value is clearly subject to error as well. In this case, however, the amount is not predicted from a single measured value of y, so the formula for the standard deviation, s_{x_E}, of the extrapolated x-value (x_E) is not the same as that in Eq. (5.9). Instead, we use:

$$s_{x_E} = \frac{s_{y/x}}{b} \left\{ \frac{1}{n} + \frac{\bar{y}^2}{b^2 \sum_i (x_i - \bar{x})^2} \right\}^{\frac{1}{2}}$$ (5.12)

Increasing the value of n again improves the precision of the estimated amount: in general, at least six points should be used in a standard-additions experiment. Moreover, the precision is improved by maximizing $\sum_i (x_i - \bar{x})^2$, so the calibration

solutions should, if possible, cover a considerable range. Confidence limits for x_E can, as always, be determined as $x_E \pm t s_{x_E}$.

Example. The silver concentration in a sample of photographic waste was determined by atomic-absorption spectrometry with the method of standard additions. The following results were obtained.

Added Ag, μg per ml of original sample solution	0	5	10	15	20	25	30
Absorbance	0.32	0.41	0.52	0.60	0.70	0.77	0.89

Determine the concentration of silver in the sample, and obtain 95% confidence limits for this concentration.

Equations (5.4) and (5.5) yield $a = 0.3218$ and $b = 0.0186$. The ratio of these figures gives the silver concentration in the test sample as 17.3 μg/ml. The confidence limits for this result can be determined with the aid of Eq. (5.12). Here $s_{y/x}$ is 0.01094, $\bar{y} = 0.6014$, and $\sum_i (x_i - \bar{x})^2 = 700$. The value of s_{x_E} is thus 0.749 and

the confidence limits are $17.3 \pm 2.57 \times 0.749$, i.e. $17.3 \pm 1.9 \ \mu$g/ml.

Although it is an elegant approach to the common problem of matrix interference effects, the method of standard additions has a number of disadvantages: it is difficult to automate, and it may use larger quantities of sample than other methods. In statistical terms its principal disadvantage is that it is an extrapolation method, and is

Fig. 5.9.   The method of standard additions.

thus less precise than interpolation techniques. In the example just given, it is easy to show that, if an unknown concentration of silver added to the sample of photographic waste yielded an absorbance value of 0.65, the concentration of the added silver would be 17.6 μg/ml, with confidence limits of 17.6 ± 1.6 μg/ml. Comparing this result with that obtained in the example, the moderately improved confidence limits are apparent. As we have seen, the confidence limits for a point on a regression line vary with the y-value, and are at a minimum when $y = \bar{y}$ (see above, Section 5.6). In practice, as shown by both this example and the example in Section 5.6, this variation in the confidence limits with change in y may not be very great, so the confidence limits are by no means bad even for a result derived from a substantial extrapolation.

5.9   USE OF REGRESSION LINES FOR COMPARING ANALYTICAL METHODS

If an analytical chemist develops a new method for the determination of a particular analyte, he will certainly wish to validate it by (amongst other techniques) applying it to a series of samples already studied by use of another reputable or standard procedure. In making such a comparison, the principal interest will be the identification of systematic errors — does the new method give results that are significantly higher or lower than the established procedure? In cases where an analysis is repeated several times over a very limited concentration range, such a comparison can be made by using the statistical tests described in a previous chapter (see Sections 3.2 and 3.3). Such procedures will not be appropriate in instrumental analyses, which are often used over large concentration ranges.

When two methods are to be compared at different analyte concentrations the procedure illustrated in Fig. 5.10 is normally adopted. One axis of a regression graph

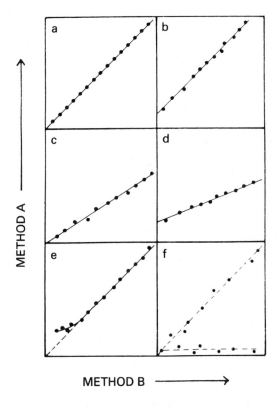

Fig. 5.10.   Use of a regression line to compare two analytical methods: (a) shows perfect agreement between the two methods for all the samples; (b)–(f) illustrate the results of various types of systematic error (see text).

is used for the results obtained by the new method, and the other axis for the results obtained by applying the reference or comparison method to the same samples. (The important question of allocation of an axis to a method is further discussed below.) Each point on the graph thus represents a single sample analysed by two separate methods. The methods of the preceding sections are then applied to calculate the slope (b), the intercept (a) and the product-moment correlation coefficient (r) of the regression line. It is clear that if each sample yields an identical result with both analytical methods the regression line will have a zero intercept, and a slope and a correlation coefficient of 1 (Fig. 5.10a). In practice, of course, this never occurs — even if systematic errors are entirely absent, random errors ensure that the two analytical procedures will not give results in exact agreement for all the samples.

Deviations from the 'ideal' situation ($a = 0$, $b = r = 1$) can occur in a number of different ways. First, it is possible that the regression line will have a slope of 1, but a non-zero intercept. That is, one method of analysis may yield a result higher or lower than the other by a fixed amount. Such an error might occur if the background signal for one of the methods was wrongly calculated (Fig. 5.10b). A second possibility is that the slope of the regression line is >1 or <1, indicating that a systematic error may be occurring in the slope of one of the individual calibration plots (Fig. 5.10c). These two errors may of course occur simultaneously (Fig. 5.10d). Further possible types of systematic error are revealed if the plot is curved (Fig. 5.10e). Speciation problems may give surprising results (Fig. 5.10f). This type of plot might arise if an analyte occurred in two chemically distinct forms, the proportions of which varied from sample to sample. One of the methods under study (here plotted on the y-axis) might detect only one form of the analyte, while the second method detected both forms.

In practice, the analyst most commonly wishes to test for an intercept differing significantly from zero, and a slope differing significantly from 1. Such tests are performed by determining the confidence limits for a and b, generally at the 95% significance level. The calculation is thus in practice very similar to that described in Section 5.5, and is most usefully performed by using a simple computer program.

Example. The level of lead in ten fruit juice samples was determined by a new potentiometric stripping analysis (PSA) method employing a glassy-carbon working electrode, and the results were compared with those obtained by using a flameless atomic-absorption spectrometry (AAS) technique. The following data were obtained (all results in μg/l.).

Sample	1	2	3	4	5	6	7	8	9	10
AAS result	35	75	75	80	125	205	205	215	240	350
PSA result	35	70	80	80	120	200	220	200	250	330

(S. Mannino, *Analyst*, 1982, **107**, 1466).

These results are plotted on a regression line (Fig. 5.11), with the AAS results assigned to the x-axis, and the PSA results to the y-axis. Using the methods of the preceding sections it may readily be shown that:

$$a = 3.87; \; b = 0.968; \; r = 0.9945$$

Further calculations show that:

$$s_{y/x} = 10.56; \; s_a = 6.64; \; s_b = 0.0357$$

and the use of the appropriate t-value for 8 degrees of freedom ($t = 2.31$) gives the 95% confidence limits for the intercept and slope as:

$$a = 3.87 \pm 15.34 \text{ and } b = 0.963 \pm 0.083$$

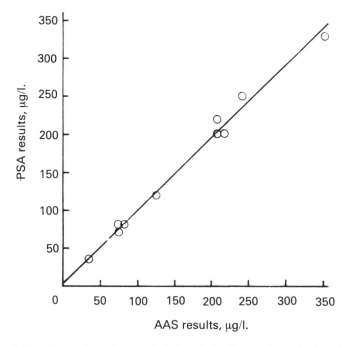

Fig. 5.11.   Comparison of two analytical methods: the plot shows the data on p. 122.

From these figures it is clear that the calculated slope and intercept do not differ significantly from the 'ideal' values of 1 and 0 respectively, and thus that there is no evidence for systematic differences between the two sets of results.

Two further points may be mentioned in connection with this example. First, inspection of the literature of analytical chemistry shows that authors frequently place great stress on the value of the correlation coefficient in such comparative studies. In the example, however, it played no direct role in establishing whether or not systematic errors had occurred. Even if the regression line had been slightly curved, the correlation coefficient might still have been close to 1 (cf. Section 5.3 above). It is thus evident that the calculation of r is of less importance in the present context than the establishment of confidence limits for the slope and the intercept. In some cases it may be found that the r value is not very close to 1, even though the slope and the intercept are not significantly different from 1 and 0 respectively. Such a result would suggest very poor precision for either one or both of the methods under study. The precision of the two methods can, of course, be determined and compared by the methods of Chapters 2 and 3. In practice it is desirable that this should be done *before* the regression line comparing the methods is plotted — the reason for this is explained below. The second point to note is that, although it is clearly desirable to compare the methods over as wide a range of concentrations as possible, it may not in practice be feasible to obtain real samples which have analyte concentrations that are evenly spaced in the range. In the example given, the fruit juice examined tended to fall into groups with fairly similar lead concentrations, so samples with lead levels between ca. 130 and 200 $\mu g/l$. and between 260 and 320 $\mu g/l$. could not be studied. This point will also be referred to below.

Although almost universally adopted in comparative studies of instrumental methods, the approach described here is open to serious theoretical objections on several grounds. First, as has been emphasized throughout this chapter, the line of regression of y on x is calculated on the assumption that the errors in the x-values are negligible — all errors are assumed to occur in the y-direction. While generally valid in the production of a calibration plot for a single analytical method, this assumption can evidently not be justified when the regression line is used for comparison purposes. In such comparisons it can be taken as certain that random errors will occur in both analytical methods, i.e. in both the x and y directions. This would suggest that the equations used to calculate the regression line itself may not be valid. Practical tests and simulations, however, show that the present simple approach does give surprisingly reliable results, provided that three conditions are fulfilled.

1. The more precise method is plotted on the x-axis: this is the reason for making preliminary investigations on the precisions of the two methods — see above.

2. A reasonable number (10 at least, as in the example above) of points are plotted in the comparison. Since the confidence limit calculations are based on $(n - 2)$ degrees of freedom, it is particularly important to avoid small values of n.

3. The experimental points should cover the concentration range of interest in a roughly uniform fashion: as we have seen, this requirement may be difficult to fulfil in comparative studies on real samples.

There is a second theoretical objection to using the line of regression of y on x, as calculated in Section 5.4 and 5.5, in the comparison of two analytical methods. This regression line assumes not only that the x-direction errors are zero: it also assumes that the error in the y-values is *constant*, i.e. it does not vary with concentration, and that all the points thus have equal weighting when the slope and intercept of the line are calculated. This assumption is obviously likely to be invalid in practice. In many analyses, the relative standard deviation (coefficient of variation) is roughly constant over a range of concentrations: the absolute error thus increases with the concentration of the analyte, rather than having the same value at all concentrations. It follows that 'unweighted' regression lines are also of very questionable validity in other situations, e.g. when applied to calibration plots for a single analytical procedure. In principle, weighted regression lines should be used instead. This concept is elaborated in the next section. Meanwhile, it can be reaffirmed that, despite the theoretical objections, unweighted regression lines provide useful information in comparative studies provided that the requirements listed above are fulfilled.

5.10   WEIGHTED REGRESSION LINES

The comments made in the previous section on conventional or unweighted regression calculations indicate that weighted regression calculations should perhaps be adopted far more frequently than is in fact the case. The calculations involved in using weighted regression methods are only a little more complicated than those of

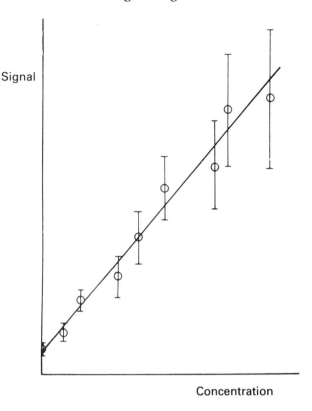

Signal

Concentration

Fig. 5.12.   The weighting of errors in a regression calculation.

the previous sections, and can easily be performed on a micro-computer, but they require additional information on the errors occurring at different concentration levels, or at least the formulation of additional assumptions about such errors. These considerations perhaps explain why weighted regression calculations are less used than they might be. In this section the application of weighted regression methods is outlined. It is assumed that the weighted regression line is to be used for the determination of a single analyte rather than for the comparison of two separate methods.

Let us consider in more detail the simple situation that arises when the error in a regression calculation is approximately proportional to the concentration of the analyte. When the errors at different points on the calibration graph are expressed by 'error bars' (Fig. 5.12) the bars thus get larger as the concentration increases. In this situation it is evident that the regression line must be calculated to give additional weight to those points where the error bars are smallest: it is more important for the calculated line to pass close to such points than to pass close to the points representing higher concentrations with the largest errors. This result is achieved by giving each point a weighting inversely proportional to the corresponding variance, s^2; this logical procedure is of general application. Thus, if the individual points are denoted by (x_1, y_1), (x_2, x_2) etc. as usual, and the corresponding standard deviations are s_1, s_2, etc., then the individual weights, w_1, w_2, etc., are given by:

$$w_i = s_i^{-2}/(\sum_i s_i^{-2}/n)$$

(5.13)

It will be seen that the weights have been scaled so that their sum is equal to the number of points on the graph: this simplifies the subsequent calculations. The slope and the intercept of the regression line are then given by:

$$b = \frac{\sum_i w_i x_i y_i - n\bar{x}_w\bar{y}_w}{\sum_i w_i x_i^2 - n\bar{x}_w^2}$$

(5.14)

and

$$a = \bar{y}_w - b\bar{x}_w$$

(5.15)

In Eq. (5.15), $\bar{x}_w$ and $\bar{y}_w$ represent the co-ordinates of the *weighted centroid*, $(\bar{x}_w, \bar{y}_w)$ through which the weighted regression line must pass. These co-ordinates are given, as expected, by $\bar{x}_w = \sum_i w_i x_i/n$ and $\bar{y}_w = \sum_i w_i y_i/n$.

Example. Calculate the unweighted and weighted regression lines for the following calibration data. For each line calculate also the concentrations of test samples with absorbances of 0.100 and 0.600.

Concentration, μg/ml	0	2	4	6	8	10
Absorbance	0.009	0.158	0.301	0.472	0.577	0.739
Standard deviation	0.001	0.004	0.010	0.013	0.017	0.022

Application of Eqs. (5.4) and 5.5) shows that the slope and intercept of the *unweighted* regression line are respectively 0.0725 and 0.0133. The concentrations corresponding to absorbances of 0.100 and 0.600 are thus readily found to be 1.20 and 8.09 μg/ml respectively.

The *weighted* regression line is a little harder to calculate: in the absence of a suitable computer program it is usual to set up a table as follows.

x_i	y_i	s_i	$1/s_i^2$	w_i	$w_i x_i$	$w_i y_i$	$w_i x_i y_i$	$w x_i^2$
0	0.009	0.001	1000000	5.535	0	0.0498	0	0
2	0.158	0.004	62500	0.346	0.692	0.0547	0.1093	1.384
4	0.301	0.010	10000	0.055	0.220	0.0166	0.0662	0.880
6	0.472	0.013	5917	0.033	0.198	0.0156	0.0935	1.188
8	0.577	0.017	3460	0.019	0.152	0.0110	0.0877	1.216
10	0.739	0.022	2066	0.011	0.110	0.0081	0.0813	1.100
		SUMS:	1083943	5.999	1.372	0.1558	0.4380	5.768

From these figures it is clear that $\bar{y}_w = 0.1558/6 = 0.0260$ and $\bar{x}_w = 1.372/6 = 0.229$. By Eq. (5.14), b is calculated from

$$b = \frac{0.438 - (6 \times 0.229 \times 0.026)}{5.768 - [6 \times (0.229)^2]} = 0.0738$$

so a is given by $0.0260 - (0.0738 \times 0.229) = 0.0091$.

These values for a and b can be used to show that absorbance values of 0.100 and 0.600 correspond to concentrations of 1.23 and 8.01 μg/ml respectively.

Careful comparison of the results of the unweighted and weighted regression calculations is very instructive. The effects of the weighting process are clear. The weighted centroid $(\bar{x}_w, \bar{y}_w)$ is much closer to the origin of the graph than the unweighted centroid $(\bar{x}, \bar{y})$, and the weighting given to the points nearer the origin — and particularly to the first point $(0, 0.009)$, which has the smallest error — ensures that the weighted regression line has an intercept very close to this point. The slope and intercept of the weighted line are remarkably similar to those of the unweighted line, however, with the result that the two methods give very similar values for the concentrations of samples having absorbances of 0.100 and 0.600. It must not be supposed that these similar values arise simply because in this example the experimental points fit a straight line very well. In practice the weighted and unweighted regression lines derived from a set of experimental data have similar slopes and intercepts even if the scatter of the points about the line is substantial.

We might thus be forgiven for thinking that weighted regression calculations have little to recommend them. They require more information (in the form of estimates of the standard deviation at various points on the graph), and are far more complex to execute, but they seem to provide data that are remarkably similar to those obtained from the much simpler unweighted regression method. Such considerations may indeed account for the widespread neglect of weighted regression calculations in practice. But an analytical chemist using instrumental methods does not employ regression calculations simply to determine the slope and intercept of the calibration plot and the concentrations of test samples. He also wants to obtain estimates of the errors or confidence limits of those concentrations and it is in this context that the weighted regression method provides much more realistic results. In Section 5.6 we used Eq. (5.9) to estimate the standard deviation (s_{x_0}) and hence the confidence intervals of a concentration calculated by using a single y-value and an unweighted regression line. Application of this equation to the data in the example above shows that the unweighted confidence limits for the solutions having absorbances of 0.100 and 0.600 are 1.20 ± 0.65 and 8.09 ± 0.63 μg/ml respectively. As in the example in Section 5.6, these confidence intervals are very similar. In the present example, however, such a result is entirely unrealistic. The experimental data show that the errors of the observed y-values increase as y itself increases, the situation expected for a method having a roughly constant relative standard deviation. We would expect that this increase in s_i with increasing y would also be reflected in the confidence limits of the determined concentrations: the confidence limits for the solution with an absorbance of 0.600 should be much greater (i.e. worse) than those for the solution with an absorbance of 0.100.

In weighted regression calculations, the standard deviation $(s_{x_{0w}})$ of a predicted concentration is given by:

$$s_{x_{0w}} = \frac{s_{(y/x)w}}{b} \left\{ \frac{1}{w_0} + \frac{1}{n} + \frac{(y_0 - \bar{y}_w)^2}{b^2(\sum_i w_i y_i^2 - n\bar{x}_w^2)} \right\}$$

(5.16)

In this equation, $s_{(y/x)w}$ is given by:

$$s_{(y/x)w} = \left\{ \frac{(\sum_i w_i y_i^2 - n\bar{y}_w^2) - b^2 (\sum_i w_i x_i^2 - n\bar{x}_w^2)}{n-2} \right\}^{\frac{1}{2}}$$

(5.17)

and w_0 is a weighting appropriate to the value of y_0. Equation (5.16) is clearly similar in form to Eq. (5.9). It confirms that points close to the origin, where the weights are highest, and points near the centroid, where $y_0 - \bar{y}_w$ is small, will have the smallest confidence limits (Fig. 5.13). The major difference between the two equations is the term $1/w_0$ in Eq. (5.16). Since w_0 falls sharply as y increases, this term ensures that the confidence limits increase with increasing y_0, as we expect.

Application of Eq. (5.16) to the data in the example above shows that the test samples with absorbance of 0.100 and 0.600 have confidence intervals for the calculated concentrations (1.23 and 8.01 μg/ml) of ± 0.12 and ± 0.72 μg/ml respectively. We note that these confidence intervals are proportional to the observed absorbances of the two solutions. In addition the confidence interval for the less concentrated of the two samples is smaller than in the unweighted regression calculation, while for the more concentrated sample the opposite is true. All these results accord much more closely with the reality of a calibration experiment than do the results of the unweighted regression calculation.

We therefore conclude that, although weighted regression calculations are more complicated to perform than their unweighted counterparts, they give much more realistic results for the confidence limits of predicted concentrations in conventional instrumental analysis. In addition, weighted regression methods may be essential when a straight line graph is obtained by algebraic transformations of an intrinsically curved plot (see below, Section 5.11). Computer programs for weighted regression calculations are now available for several types of personal computer, and should encourage the more widespread use of this method.

5.11 CURVILINEAR REGRESSION — INTRODUCTION

Thus far, our discussion of calibration methods has been confined to experiments where we can assume that the instrument response is proportional to the analyte concentration. This assumption is often valid, because analytical chemists have

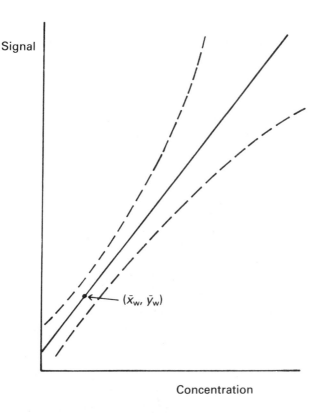

Fig. 5.13.   General form of the confidence limits for a concentration determined by using a
weighted regression line.

always — until quite recently — favoured such methods in view of the complexity of curve-fitting calculations, and have thus taken precautions to ensure linearity of response over a wide concentration range. Examples of such precautions include the control of the emission line-width of a hollow-cathode lamp in atomic-absorption spectrometry, and the size and positioning of the sample cuvette to minimize inner filter effects in molecular fluorescence spectrometry. Despite such endeavours, however, many analytical methods give curved calibration plots over the concentration range of interest. Particularly common is the situation where the calibration plot is linear (or approximately so) at low analyte concentrations, but becomes curved at higher analyte levels. In other cases, e.g. immunoassays, the calibration plot is obviously curved at all concentrations. When calibration plots are curved we shall still need answers to the questions listed in Section 5.2, but the questions will pose rather more formidable statistical problems than do linear calibration experiments.

The first question to be examined is — how do we detect curvature in a calibration plot? That is, how do we distinguish between a plot which is best fitted by a straight line, and one which is best fitted by a gentle curve? Since the degree of curvature may be small, and/or occur over only part of the plot, this is clearly not a trivial question. Moreover, despite its widespread use for testing the goodness of fit of linear graphs, the product-moment correlation coefficient (r) is of little value in testing for curvature: we have seen (Section 5.3) that lines with obvious curvature may still give very high r values. An analyst would naturally hope that any test for curvature could be applied fairly easily in routine work without extensive calculations. Several such tests are available, based on the use of the y-residuals on the calibration plot.

As we have seen (Section 5.5) a y-residual, $y_i - \hat{y}_i$, represents the difference between an experimental value of y and the $\hat{y}$ value calculated from the regression line at the same value of x. If a linear calibration plot is appropriate, and if the random errors in the y-values are normally distributed, the residuals themselves should be normally distributed about the value of zero. If this turns out not to be true in practice, then (neglecting the possibility that the experimental errors themselves are not normally distributed) we must suspect that the fitted regression line is not of the correct type. In the worked example given in Section 5.5, the y-residuals were shown to be $+0.58$, -0.38, -0.24, -0.50, $+0.34$, $+0.18$, and $+0.02$. It is clear that these values sum to zero (allowing for possible rounding errors, this must always be true), and are approximately symmetrically distributed about 0. Although it is impossible to be certain, especially with small numbers of data points, that these residuals are normally distributed, there is certainly no contrary evidence in this case, i.e. no evidence to support a non-linear calibration plot.

A second test suggests itself on inspection of the signs of the residuals given above. As we move along the calibration plot, i.e. as x increases, positive and negative residuals will be expected to occur in random order if the data are well fitted by a straight line. If, in contrast, we attempt to fit a straight line to a series of points that actually lie on a smooth curve, then the signs of the residuals will no longer have a random order, but will occur in *sequences* of positive and negative values. Examining again the residuals given above, we find that the order of signs is $+ - - - - + + +$. To test whether these sequences of $+$ and $-$ residuals indicate the need for a non-linear regression line, we need to know the probability that such an order could occur by chance; such calculations are described in the next chapter. Unfortunately the small number of data points makes it quite likely that these and other sequences could indeed occur by chance, so any conclusions drawn must be treated with great caution. The choice between straight line and curvilinear regression methods is probably best made by using the curve-fitting techniques outlined below.

In the situation where a calibration plot is linear over part of its range and curved elsewhere, it is of great importance for the analytical chemist to establish the range over which linearity can be assumed. Possible approaches to this problem are outlined in the following example.

Example. Investigate the linear calibration range of the following fluorescence experiment.

Fluorescence intensity	0.1	8.0	15.7	24.2	31.5	33.0
Concentration, μg/ml	0	2	4	6	8	10

Inspection of the data shows that the part of the graph near the origin corresponds rather closely to a straight line with a near-zero intercept and a slope of about 4. The fluorescence of the 10-μg/ml standard solution is clearly lower than would be expected on this basis, and there is some possibility that the departure from linearity has also affected the fluorescence of the 8-μg/ml standard. We first apply (unweighted) linear regression calculations to all the data. Application of the methods of Sections 5.3 and 5.4 gives the results $a = 1.357$, $b = 3.479$ and $r = 0.9878$. Again we recall that the high value for r may be deceptive, though it may be used in a comparative sense (see below). The y-residuals are found to be -1.257, -0.314, $+0.429$, $+1.971$, $+2.314$, and -3.143, with the sum of squares of the residuals equal to 20.981. The trend in the values of the residuals suggests that the last value in the table is probably outside the linear range.

We confirm this suspicion by applying the linear regression equations to the first five points only. This gives $a = 0.100$, $b = 3.950$ and $r = 0.9998$. The slope and intercept are much closer to the values expected for the part of the graph closest to the origin, and the r value is higher than in the first calculation. The residuals of the first five points from this second regression equation are $0, 0, -0.2, +0.4$ and -0.2, with a sum of squares of only 0.24. Use of the second regression equation shows that the fluorescence expected from a 10-μg/ml standard is 39.6, i.e. the residual is -6.6. Use of a t-test (Chapter 3) would show that this last residual is significantly greater than the average of the other residuals: alternatively a test could be applied (Section 3.6) to demonstrate that it is an 'outlier' amongst the residuals. In this example, such calculations are hardly necessary: the enormous residual for the last point, coupled with the very low residuals for the other five points and the greatly reduced sum of squares, confirms that the linear range of the method does not extend as far as 10 μg/ml.

Having established that the last data point can be excluded from the linear range, we can repeat the process to study the point (8, 31.5). We do this by calculating the regression line for only the first four points in the table, with the results $a = 0$, $b = 4.00$, $r = 0.9998$. The correlation coefficient value suggests that this line is about as good a fit of the points as the previous one, in which five points were used. The residuals for this third calculation are $+0.1$, 0, -0.3, and $+0.2$, with a sum of squares of 0.14. With this calibration line the y-residual for the 8-μg/ml solution is -0.5: this value is larger than the other residuals but probably not by a significant amount. It can thus be concluded that it is reasonably safe to include the point (8, 31.5) within the linear range of the method. In making a marginal decision of this kind, the analytical chemist will take into account the accuracy required in his results, and the reduced value of a method for which the calibration range is very short. The calculations described above are summarized in Fig. 5.14.

Once a decision has been taken that a set of calibration points cannot be satisfactorily fitted by a straight line, the analyst can play one further card before resigning himself to the complexities of curvilinear regression calculations. He may be able to **transform** the data so that a non-linear relationship is changed into a linear

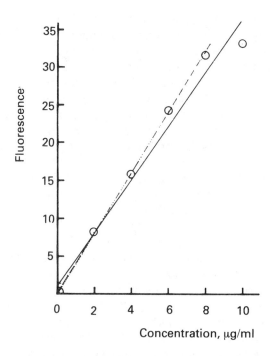

Fig. 5.14.   Curvilinear regression: identification of the linear range. The data on p. 130 are used; the unweighted linear regression lines through all the points (———), through the first five points only (– – – – –), and through the first four points only (.....) are shown.

one. Such transformations are regularly applied to the results of certain analytical methods. For example, modern software packages for the interpretation of immunoassay data frequently offer a choice of transformations: commonly used methods involve plotting $\log y$ and/or $\log x$ instead of y and x, or the use of logit functions (logit $x = \ln[x/(1-x)]$). It is important to note that the transformations may also affect the nature of the errors at different points on the calibration plot. Suppose, for example, that in a set of measurements of the form $y = px^q$, the magnitudes of the random errors in y are independent of x. Transformation of the data into linear form by taking logarithms will obviously produce data in which the errors in $\log y$ are *not* independent of $\log x$. In this case, and in any other instance where the expected form of the equation is known from theoretical considerations or from longstanding experience, it is possible to apply *weighted* regression equations (Section 5.10) to the transformed data. It may be shown that, if data of the general form $y = f(x)$ are transformed into the linear equation $Y = BX + A$, the weighting factor, w_i, used in equations (5.14)–(5.17) is obtained from the relationship:

$$w_i = (1/[dY_i/dy_i])^2 \qquad\qquad (5.18)$$

In the present example, $y = px^q$, so $Y = \ln y$, and $dY_i/dy_i = d \ln y_i/dy_i = 1/y_i$. Thus

$w_i = y_i^2$. Unfortunately, there are not many cases in analytical chemistry where the exact mathematical form of a non-linear regression equation is known with certainty, so this approach is of restricted value.

It should also be noted that, in contrast to the situation described in the previous paragraph, results may be transformed to produce data that can be treated by *unweighted* methods. Data of the form $y = bx$ with y-direction errors strongly dependent on x are sometimes subjected to a log-log transformation: the errors in log y then vary less seriously with log x, so the transformed data can reasonably be studied by unweighted regression equations.

5.12   CURVE FITTING

In view of the difficulties that arise from transforming the data, and the increasing ease with which curves can be calculated to fit a set of calibration points, curvilinear regression methods are now relatively common in analytical chemistry. It is important to realize that the curved calibration plots encountered in practice often arise from the superposition of two or more physical or chemical phenomena. Thus in molecular fluorescence spectrometry, signal vs. concentration plots will often be approximately linear in very dilute solution, but will show increasing (negative) curvature at higher concentrations because of (a) optical artifacts (inner filter effects), (b) molecular interactions (e.g. quenching, excimer formation), and (c) the failure of the algebraic assumptions on which a linear plot is predicted. Effects (a)–(c) are independent of one another, so many curves of different shape may appear in practice. It is for reasons of this kind that calibration curves of a known and predictable form are so rarely encountered in analytical work (see above), so the analyst has little *a priori* guidance on which of the many types of equation that generate curved plots should be used to fit the calibration data. In practice, much the most common strategy is to fit a curve which is a polynomial in x, i.e. $y = a + bx + cx^2 + dx^3 + \ldots$. The mathematical problems to be solved are then (i) how many terms should be included in the polynomial, and (ii) what values must be assigned to the coefficients a, b etc? Computer software packages which address these problems are normally iterative: they fit first a straight line, then a quadratic curve, then a cubic curve, and so on, to the data, and present to the user the information needed to decide which of these equations is the most suitable. Before studying in more detail how this decision is made, it is worth noting that, in practice, quadratic or cubic equations are often entirely adequate to provide a good fit to the data: polynomials with many terms are almost certainly physically meaningless and do not significantly improve the analytical results. In any case, if the graph has n calibration points, the largest polynomial permissible is that of order $(n - 1)$.

How then do we decide whether (for example) a quadratic or a cubic curve is the best fit to a calibration data set? Just as in the case of straight-line plotting it seems reasonable to use the 'least-squares' criterion to decide the quality of the fit, i.e. we seek to minimize the sum of the squares of the y-residuals. This approach allows us to use simple analysis of variance (ANOVA, see Chapter 3) calculations to evaluate the quality of successive polynomials. The ANOVA calculation assumes (see Section 5.2 above) that only y-direction random errors occur, and operates as follows. If the individual calibration points have values $y_1, y_2, \ldots y_i, \ldots y_n$, and mean value $\bar{y}$ as usual, then the term $\Sigma[y_i - \bar{y}]^2$ is known as the 'sum of squares (SS) about $\bar{y}$'. It may be shown that it is equal to the sum of two other terms, the 'sum of squares due to regression' ($\Sigma[\hat{y}_i - \bar{y}]^2$), and the 'sum of squares about regression' ($\Sigma[y_i - \hat{y}_i]^2$). The sum of squares due to regression is evidently that part of $\Sigma[y_i - \bar{y}]^2$ which is due to the successful fitting of an equation to the data points — it should therefore be as large as possible. Conversely, the sum of squares about regression is clearly equal to the sum of squares of the y-residuals (the 'residual sum of squares'), which should be as small as possible if the curve is a good fit to the data points. We can thus define R^2, the **coefficient of determination**, as

$$R^2 = \frac{\text{SS due to regression}}{\text{Total SS}} = 1 - \frac{\text{Residual SS}}{\text{Total SS}} \tag{5.19}$$

R^2 (sometimes also called the multiple correlation coefficient) clearly varies between 0 and 1, the latter representing a perfect fit of a curve to a set of data points. It can be shown that, in the case of a straight-line graph, $R^2 = r^2$, the square of the product-moment correlation coefficient (Section 5.3). The term **coefficient of non-determination** is sometimes applied to $1 - R^2$. In summary, the ANOVA calculations provided by curve-fitting programs divide the total sum of squares of the y_i-values about $\bar{y}$ (this sum is of course a constant for any given data set) into two components, whose contributions to the total vary according to the goodness of the fit, given by R^2.

In deciding which polynomial provides the best fit to a set of calibration points lying on a curve, it would thus seem that we have only to calculate R^2 values for the straight-line, quadratic, cubic etc. equations, and cease our search when R^2 no longer increases. Unfortunately it turns out that the addition of another term to the polynomial *always* increases the value of R^2, even if only by a small amount. Many programs thus provide an additional or alternative statistic, usually known 'as adjusted R^2' or R'^2, which utilizes mean square (MS) values rather than sums of squares, and is given by:

$$R'^2 = 1 - \frac{\text{residual MS}}{\text{total MS}} \tag{5.20}$$

The use of R'^2 takes into account that the number of residual degrees of freedom in the polynomial regression (given by $[n - k - 1]$ where k is the number of terms in the regression equation containing a function of x) changes as the order of the polynomial changes (see below). As the following example shows, R'^2 is always smaller than R^2.

Example. In an instrumental analysis the following data were obtained (arbitrary units):

Concentration	0	1	2	3	4	5	6	7	8	9	10
Signal		0.2	3.6	7.5	11.5	15.0	17.0	20.4	22.7	25.9	27.6	30.2

Fit a suitable polynomial to these results, and use it to estimate the concentrations corresponding to signal of 5, 16 and 27 units.

Even a casual examination of the data suggests that the calibration plot should be a curve, but it is instructive nonetheless to calculate the least-squares straight line through the points using the method described in Section 5.4. This line turns out to have the equation $y = 2.991x + 1.555$. When the ANOVA table for the data (see Section 3.10) is calculated it is found to have the following form:

Source of variation	Sum of squares	D.f.	Mean square
Regression	984.009	1	984.009
Residual	9.500	9	1.056
Total	993.509	10	99.351

It may be shown that in such a table the number of degrees of freedom (d.f.) for the variation due to regression is equal to k, the number of terms in the regression equation containing x, x^2, etc. For a straight line, k is clearly 1, as we have only a term containing x itself. As we have only one constraint in the calculation (viz. that the sum of the residuals is zero, see above), the total number of degrees of freedom is $n - 1$. Thus the number of degrees of freedom assigned to the residuals is $(n - k - 1) = (n - 2)$ in this case. (We have already utilized this result in the calculations described in Sections 5.5, 5.6, 5.8 and 5.9.) From the data in the ANOVA table it is clear that R^2 is given by $984.009/993.509 = 0.99044$, i.e. 99.044%. It might seem that an equation which explains over 99% of the relationship between x and y is very satisfactory but, just as is the case with the correlation coefficient, r, we must use great caution in interpreting absolute values of R^2: it will soon become apparent that a quadratic curve provides a much better fit for the data. We can also calculate the adjusted R^2 value from equation (5.20): it is given by $(1 - [1.056/99.351]) = 0.98937$, i.e. 98.937%.

We have already noted that an examination of the residuals usually provides valuable information on the success of a calibration equation. Many computer programs provide output of the following kind:

x	y_i	$\hat{y}_i$	y-residual
0	0.2	1.6	−1.4
1	3.6	4.5	−0.9
2	7.5	7.5	0
3	11.5	10.5	1.0
4	15.0	13.5	1.5
5	17.0	16.5	0.5
6	20.4	19.5	0.9
7	22.7	22.5	0.2
8	25.9	25.5	0.4
9	27.6	28.5	−0.9
10	30.2	31.5	−1.3

In this table, the numbers in the two right-hand columns have been rounded to one decimal place for simplicity. The trend in the signs and magnitudes of the residuals, which are negative at low x-values, rise to a positive maximum, and then return to negative values, is immediately apparent, and a sure sign that a straight line is not a suitable fit for the data.

When the data are fitted by a curve of quadratic form, the equation turns out to be $y = 0.086 + 3.970x - 0.098x^2$, and the ANOVA table takes the form:

Source of variation	Sum of squares	D.f.	Mean square
Regression	992.233	2	496.116
Residual	1.276	8	0.160
Total	993.509	10	99.351

Note that the number of degrees of freedom for the regression and residual sources of variation has now changed in accordance with the rules described above, but that the total variation is naturally the same as in the first ANOVA table. Here we can readily show that R^2 is 992.233/993.509 = 0.99872, i.e. 99.872%. This figure is noticeably higher than the value of 99.044% obtained from the linear plot, and the adjusted R^2 value is also higher at $(1 - [0.160/99.351]) = 0.99839$, i.e. 99.839%. When the y-residuals are calculated, their signs (in increasing order of x-values) are $+ - - + + - + - + - +$: there is no obvious trend here, so it is clear that on all grounds we must prefer the quadratic over the linear fit.

Lastly we repeat the calculation for a cubit fit. Here, the best-fit equation is $y = -0.040 + 4.170x - 0.150x^2 + 0.0035x^3$. The cubic coefficient is very small indeed, so it is questionable whether this equation is a significantly better fit than the quadratic one. The R^2 value is, inevitably, slightly higher than that for the quadratic curve (99.879% compared with 99.872%), but the value of adjusted R^2 is slightly *lower* than the quadratic value at 99.827%. The order of the signs of the residuals is the same as in the quadratic fit. As there is no value in including unnecessary terms, we can be confident that a quadratic fit is satisfactory in this case.

When the above equations are used to estimate the concentrations corresponding to instrument signals of 5, 16 and 27 units, the results (x-values in arbitrary units) are:

	Linear	Quadratic	Cubic
$y = 5$	1.15	1.28	1.27
$y = 16$	4.83	4.51	4.50
$y = 27$	8.51	8.61	8.62

As expected, the differences between the concentrations calculated from the quadratic and cubic equations are insignificant, so the quadratic equation is used for simplicity.

We noted earlier in this section that non-linear calibration graphs often result from the simultaneous occurrence of a number of physiochemical and/or mathematical-phenomena, so it is sensible to assume that no single mathematical function may be able to describe the calibration curve entirely satisfactorily. It thus seems logical to try to fit the points to a curve which consists of several linked sections whose mathematical form may be different. This is the approach now used with increasing frequency though the application of **spline functions**. **Cubic splines** are most commonly used in practice, i.e. the final curve is made up of a series of linked sections of cubic form. These sections must clearly form a continuous curve at their junctions ('knots'), so the first two derivatives of each curve at any knot must be identical. A variety of methods have been used for estimating both the number of knots and the equations of the curves joining them: these techniques are too advanced to be considered in detail here, but it should be noted that many commer-cially available statistics software packages now provide such facilities, and that the spline function approach has been applied successfully to a variety of analytical methods, including gas-liquid chromatography, competitive binding immunoassays, and atomic-emission spectrometry.

It is legitimate to ask whether, in the case of a calibration plot whose curvature is not too severe, we might not take the spline idea to its simplest conclusion, and plot the curve as a series of straight lines joining successive points. This method is of course entirely non-rigorous, and would not provide any information on the precision with which x-values can be determined. However, its value as a simple initial data analysis (IDA) method (see Chapter 6) is indicated by applying it to the data in the above example. For y-values of 5, 16 and 27 this method of linear interpolation between successive points gives x-values of 1.36, 4.50 and 8.65 units respectively. Comparison with the above table shows that these results, especially the last two, would be quite acceptable for many purposes.

5.13 OUTLIERS IN REGRESSION

In this section we return to a problem already discussed in Chapter 3, viz, the occurrence of outliers in our data. These anomalous results inevitably arise in calibration experiments, just as they occur in replicate measurements, but it is rather harder to deal with them in regression statistics, and researchers continue to develop new approaches in this area. One difficulty is that, although the individual y_i-values in a calibration experiment are assumed to be independent of one another, the residuals $(y_1 - \hat{y}_1$ etc.) are *not* independent of one another, as their sum is always zero. It is therefore not normally permissible to take the residuals as if they were a conventional set of replicate measurements, and apply (for example) a Q-test to identify any outliers. (If the number of y_i-values is large — a condition not generally met in analytical work — this prohibition can be relaxed).

How then do we identify outliers in a typical calibration experiment? First we note that, in cases where an obvious error such as a transcription mistake or an instrument malfunction has occurred, it is natural and permissible to reject the resulting measurement (and, if possible, to repeat it). If there are suspect measure-ments for which there are no obvious sources of error, we must clearly return to a study of the residuals. Most computer programs handling regression data provide 'residual diagnostics' routines. Some of these are simple, including plots of the individual residuals against y_i-values (Fig. 5.15). Such plots would normally be expected to show that, if the correct calibration model has been used, the residuals remain roughly uniform in size as y_i increases, and normally distributed about zero. The figure also illustrates cases where the y-direction errors increase with y_i (Section 5.10), and where the wrong regression equation has been used (Sections 5.11 and 5.12). Similarly, the y-residuals can be plotted against time if instrument drift or any other time-dependent effect is suspected. These plots show up suspect values very clearly, but do not provide criteria which can be immediately used to reject or accept them. Moreover, they are of limited value in many analytical chemistry experiments, where the number of calibration points is often small.

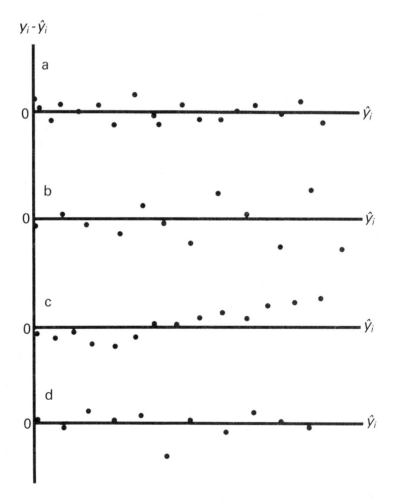

Fig 5.15.   Residual plots in regression diagnostics: (a) a satisfactory distribution of residuals; (b) the residuals tend to grow as y_i grows, suggesting that a weighted regression plot would be suitable; (c) the residuals show a trend, first becoming more negative, then passing through zero, and then becoming more positive as y_i increases, suggesting that a (different) curve should be plotted; and (d) a satisfactory plot, except that y_6 might be an outlier.

Some simple numerical criteria have been used in computer software to identify possible outliers. Some packages 'flag' calibration points where the y-residual is more than twice (or some other multiple of) the value of $s_{y/x}$. Several more advanced methods have been developed, of which the best known is the estimation for each point of 'Cook's distance', first proposed in 1977. This statistic is provided routinely by several advanced suites of statistical software, though a full appreciation of its significance requires a knowledge of matrix algebra. Cook's distance is an example of an *influence function*, i.e. it measures the effect that rejecting the calibration point in question would have on the regression coefficients a and b.

Finally we note that, just as in the treatment of outliers in replicate measurements, non-parametric and robust methods can be very effective in handling outliers in regression: robust regression methods have proved particularly popular in recent years. These topics are covered in the next chapter.

BIBLIOGRAPHY

F. S. Acton, *Analysis of Straight Line Data*, Dover, New York, 1966. A well-written and comprehensive treatment, with plenty of examples, including chemical ones. Despite its title, it also gives some attention to curvilinear regression.

R. Caulcutt and R. Boddy, *Statistics for Analytical Chemists*, Chapman & Hall, London, 1983. An entirely pragmatic approach, lacking statistical theory but full of commonsense examples. Oriented towards industrial analytical chemistry.

O. L. Davies and P. L. Goldsmith, *Statistical Methods in Research and Production*, Longmans, London, 1982. A fairly full treatment of regression and correlation methods, with considerable emphasis on advanced concepts such as curvilinear regression and multiple regression problems.

N. R. Draper and H. Smith, *Applied Regression Analysis*, Wiley, New York, 1966. An established work in this field, with a comprehensive coverage.

H. Kaiser and A. C. Menzies, *The Limit of Detection of a Complete Analytical Procedure*, Hilger, London, 1968. An interesting discussion of detection limits.

G. W. Snedecor and W. G. Cochran, *Statistical Methods*, Iowa State University, 1967. Gives an excellent general account of regression and correlation procedures.

C. Liteanu and I. Rîcă, *Statistical Theory and Methodology of Trace Analysis*, Horwood, Chichester, 1980. Wide coverage of trace analysis methods.

D. G. Kleinbaum, L. L. Kupper, and K. E. Muller, *Applied Regression Analysis and Other Multivariable Methods*, 2nd edn., PWS-Kent, Boston, 1988. Another classic in its field, with a good treatment of regression diagnostics.

A. L. Edwards, *An Introduction to Linear Regression and Correlation*, 2nd edn., W. H. Freeman, New York, 1984. Clearly written introductory treatment, with a good introduction to matrix algebra.

H. Mark, *Principles and Practice of Spectroscopic Calibration*, John Wiley, New York, 1991. Strong emphasis on the applications of near-IR spectrometry.

EXERCISES

1. In a laboratory containing polarographic equipment six samples of dust were taken at various distances from the polarograph and the mercury content of each sample was determined. The following results were obtained.

Distance from polarograph, m:	1.4	3.8	7.5	10.2	11.7	15.0
Mercury concentration, ng/g:	2.4	2.5	1.3	1.3	0.7	1.2

Examine the possibility that the mercury contamination arose from the polarograph.

2. The response of a colorimetric test for glucose was checked with the aid of standard glucose solutions. Determine the correlation coefficient from the following data, and comment on the result.

Glucose concentration, mM:	0	2	4	6	8	10
Absorbance:	0.002	0.150	0.294	0.434	0.570	0.704

3. The following results were obtained when each of a series of standard silver solutions was analysed by flame atomic-absorption spectrometry.

Concentration, ng/ml:	0	5	10	15	20	25	30
Absorbance:	0.003	0.127	0.251	0.390	0.498	0.625	0.763

Determine the slope and intercept of the calibration plot, along with their confidence limits.

4. Using the data of exercise 3, estimate the confidence limits for the silver concentrations in (a) a sample giving an absorbance of 0.456 in a single determination, and (b) a sample giving absorbance values of 0.308, 0.314, 0.347, and 0.312 in four separate analyses.

5. Estimate the limit of detection of the silver analysis from the data in exercise 3.

6. The gold content of a concentrated sea-water sample was determined by using atomic-absorption spectrometry with the method of standard additions. The results obtained were as follows.

Gold added, ng per ml of concentrated sample:	0	10	20	30	40	50	60	70
Absorbance:	0.257	0.314	0.364	0.413	0.468	0.528	0.574	0.635

Estimate the concentration of the gold in the concentrated sea-water, and determine confidence limits for this concentration.

7. The fluorescence of each of a series of acidic solutions of quinine was determined 5 times; the results are given below.

Concentration, ng/ml:	0	10	20	30	40	50
Fluorescence intensity	4	22	44	60	75	104
(arbitrary units).	3	20	46	63	81	109
	4	21	45	60	79	107
	5	22	44	63	78	101
	4	21	44	63	77	105

Determine the slope and intercept of (a) the unweighted regression line, and (b) the weighted regression line. Calculate, using both regression lines, the confidence limits for the concentrations of solutions with fluorescence intensities of 15 and 90 units.

8. An ion-selective electrode (ISE) determination of sulphide from sulphate-reducing bacteria was compared with a gravimetric determination. The results obtained were expressed in milligrams of sulphide.

Sample:	1	2	3	4	5	6	7	8	9	10
Sulphide (ISE method):	108	12	152	3	106	11	128	12	160	128
Sulphide (gravimetry):	105	16	113	0	108	11	141	11	182	118

Comment on the suitability of the ISE method for this sulphide determination. (I. K. Al-Hitti, G. J. Moody and J. D. R. Thomas, *Analyst,* 1983, **108,** 43).

9. In the determination of lead in aqueous solution by electrochemical atomic-absorption spectrometry with graphite-probe atomization, the following results were obtained.

Lead concentration, ng/ml:	10	25	50	100	200	300
Absorbance:	0.05	0.17	0.32	0.60	1.07	1.40

Investigate the linear calibration range of this experiment. (Based on S. K. Giri, C. K. Shields, D. Littlejohn and J. M. Ottaway, *Analyst,* 1983, **108,** 244).

6

Non-parametric and robust methods

6.1 INTRODUCTION

The statistical tests developed in the previous chapters have all assumed that the data being examined follow the normal (Gaussian) distribution. Some support for this assumption is provided by the central limit theorem, which shows that the sampling distribution of the mean may be approximately normal even if the parent population has quite a different distribution. However, this approximation improves as the size of the sample increases, and the value of the theorem is diminished by the very small data sets (often only three or four readings) frequently used in analytical work.

There are several further reasons for an interest in methods that do not require the assumption of normally distributed data. Some sets of data that are of interest to analytical chemists certainly have different distributions. For example (see Chapter 2) the concentrations of antibody in the blood sera of a group of different people can be expressed approximately as a log-normal distribution: such results are often obtained when a particular measurement is made on each member of a group of human or animal subjects. More interestingly, there is growing evidence that, even when repeated measurements are made on a *single* test material, the distribution of the results may sometimes be symmetrical but not normal: the data include more results than expected which are distant from the mean. Such **heavy-tailed** distributions may be regarded as arising from a normal distribution with the addition of outliers (see Chapter 3) arising from gross errors. Alternatively heavy-tailed data may arise from the superposition of two normal distributions with the same mean value, but with one distribution having a much larger standard deviation than the other: this could arise if, for example, the measurements were made by more than one individual, or by using more than one piece of equipment.

This chapter introduces two groups of statistical tests for handling data which may not be normally distributed. Methods which make no assumptions about the shape of the distribution from which the data are taken are called **non-parametric** or **distribution-free** methods. Many of them have the further advantage of greatly simplified calculations: with small data sets some of the tests can be performed mentally. The second group of methods, which has grown rapidly in use in recent years, is based on the belief that the underlying population distribution may indeed be normal (or have some other well-defined form), but with the addition of data such as outliers that may distort this distribution. These **robust** techniques will naturally be appropriate in the cases of heavy-tailed distributions described above. They differ from non-parametric methods in another respect: they often involve iterative calculations that would be lengthy or complex without a computer, and their rise in popularity certainly owes much to the universal availability of personal computers.

6.2   THE MEDIAN: INITIAL DATA ANALYSIS

In previous chapters we have used the arithmetic mean or average as the 'measure of central tendency' of a set of results. This is logical enough when the (symmetrical) normal distribution is assumed, but in non-parametric statistics, the **median** is usually used instead. In order to calculate the median of n observations, we arrange them in ascending order (in the unlikely event that n is very large, this sorting process can be performed very quickly by simple programs available for most microcomputers). The median is then the value of the $\frac{1}{2}(n + 1)$th observation if n is odd: and the average of the $\frac{1}{2}n$th and the $(\frac{1}{2}n + 1)$th observations if n is even. Determining the median of a set of experimental results thus normally requires little or no calculation. Moreover, in many cases it may be a more realistic measure of central tendency than the arithmetic mean.

Example.   Determine the mean and the median for the following four titration values.

25.01, 25.04, 25.06, 25.21 ml

It is easy to calculate that the mean of these four observations is 25.08 ml, and that the median — in this case the average of the 2nd and 3rd values, the observations already being in numerical order — is 25.05 ml. The mean is greater than any of the three closely-grouped values (25.01, 25.04 and 25.06 ml) and may thus be a less realistic measure of central tendency than the median. Instead of calculating the median we could use the methods of Chapter 3 to test the value 25.21 as a possible outlier, and determine the mean according to the result obtained, but this approach involves extra calculation and *assumes* that the data come from a normal population.

This simple example illustrates one valuable property of the median: it is unaffected by outlying values. Confidence limits (cf. Chapter 2) for the median can be estimated with the aid of the binomial distribution. This calculation can be performed even when the number of measurements is small, but is not likely to be required in analytical chemistry, where the median is generally used only as a rapid measure of an average. The reader is referred to the bibliography for further information.

In non-parametric statistics the usual measure of dispersion (replacing the standard deviation) is the **interquartile range**. As we have seen, the median divides the sample of measurements into two equal halves: if each of these halves is further divided into two the points of division are called the **upper** and **lower quartiles**. Several different conventions are used in making this calculation, and the interested reader should again consult the bibliography. The interquartile range is not widely used in analytical work, but various statistical tests can be performed on it.

The median and the interquartile range of a set of measurements are just two of the statistics which feature strongly in **initial data analysis** (IDA), often also called *exploratory data analysis* (EDA). This is an aspect of statistics which has grown rapidly in popularity in recent years. One reason for this is, yet again, the ability of modern computers and dedicated software to present data almost instantly in a wide range of graphical formats; as we shall see, such pictorial representations form an important element of IDA. A second element in the rising importance of IDA is the increasing acceptance of statistics as a practical and pragmatic subject, not necessarily restricted to the use of techniques whose theoretical soundness is unquestioned: some IDA methods seem almost crude in their principles, but have nonetheless proved most valuable.

The main advantage of IDA methods is their ability to indicate which (if any) further statistical methods are most appropriate to a given data set. Several simple presentation techniques are of immediate help. We have already noted (see Chapters 1 and 3) the use of **dot-plots** in the illustration of small data sets. These plots help in the visual identification of outliers and other unusual features of the data. The following example further illustrates their value.

Example. In an experiment to determine whether Pb^{2+} ions interfered with the enzymatic determination of glucose in various foodstuffs, nine food materials were treated with a 0.1 mM solution of Pb(II), while four other (the control group) were left untreated. The rates (arbitrary units) of the enzyme catalysed reaction were then measured for each food, and corrected for the different amounts of glucose known to be present. The results were:

Treated foods	21	1	4	26	2	27	11	24	21
Controls	22	22	32	23					

Comment on these data.

Written out in two rows as above, the data do not convey much immediate meaning, and an unthinking analyst might proceed straight away to perform a *t*-test (Chapter 3), or perhaps one of the non-parametric tests described below, to see if the two sets of results are significantly different. When, however, the data are presented as two dot plots, or perhaps a single plot with the two sets of results given separate symbols (the reader is recommended to sketch such a plot), it is immediately apparent that the results, while interesting, are so inconclusive that not much can be deduced from them *without the addition of further measurements.* The medians of the two sets of data are similar: 21 for the treated foods, and 22.5 for the controls. But the range of reaction rates for the Pb(II) treated materials is enormous, with the results apparently falling into at least two groups: five of the foods seem not to be affected by the lead (perhaps because in these cases Pb^{2+} is complexed by components other than the enzyme in question), while three others show a large inhibition effect (i.e.

the reaction rate is much reduced), and another lies somewhere in between these two extremes. There is the further problem that one of the control group results is distinctly different from the rest, and might be considered as an outlier (Chapter 3). In these circumstances it seems most unlikely that a conventional significance test will reveal chemically useful information: the use of the simplest IDA method has guided us away from thoughtless significance testing and (as so often happens) towards more experimental measurements.

Another simple data representation technique, of greater value when larger samples are studied, is the **box and whisker plot**. In its normal form such a diagram consists of a rectangle (the box) with two lines (the whiskers) extending from opposite edges of the box, and a further line in the box, crossing it parallel to the same edges. The ends of the whiskers indicate the range of the data, the edges of the box from which the whiskers protrude represent the upper and lower quartiles, and the line crossing the box represents the median of the data. The box and whisker plot, accompanied by a numerical scale, is thus a graphical representation of the **five-number summary**, i.e. the data set is described by its extremes, its lower and upper quartiles, and its median. The plot shows at a glance the spread and the symmetry of the data. Some computer programs enhance the data presentation by identifying possible outliers separately. In such cases outliers are often defined as data points which are lower than the lower quartile, or higher than the upper quartile, by more than 1.5 times the interquartile range: the whiskers then only extend to these upper and lower limits or **fences** and outlying data are shown as separate points.

Example. The levels of a blood plasma protein in 20 men and 20 women (mg/100 ml) were found to be:

```
Men      3  2  1  4  3  2  9  13 11  3 18  2  4  6  2  1  8  5  1 14
Women 6  5  2  1  7  2  2  11  2  1  1  3 11  3  2  3  2  1  4  8
```

What information can be gained about any differences between the levels of this protein in men and women?

As in the previous example, the data as presented convey very little, but the use of two box and whisker plots is very revealing. The five-number summaries are:

	Min.	Lower Quartile	Median	Upper Quartile	Max.
Men	1	2	3.5	9.5	18
Women	1	2	2.5	5.5	11

It is left as a simple sketching exercise for the reader to show that (a) the distributions are very skewed in both men and women, so statistical methods that assume a normal distribution are not appropriate (as we have seen this is often true when a single measurement is made on a number of *different* subjects, particularly when the latter are living organisms); (b) the median concentrations for men and women are similar; and (c) the range of values is considerably greater for men than for women. These conclusions suggest that we might proceed to apply the Siegel–Tukey test (see below, Section 6.6) or another non-parametric test to see whether the greater variation in protein levels amongst men is significant.

In summary, IDA methods are simple, readily handled by personal computers, and most valuable in indicating features of the data not apparent on initial inspection, in suggesting the most suitable significance tests or other statistical procedures to be adopted in further work, and sometimes even in suggesting that statistics has no further role to play until more data are obtained. They can of course be extended to the area of calibration and other regression techniques: the very crude method of plotting a curved calibration graph suggested at the end of the previous chapter can be regarded as an IDA approach. Numerous techniques are described in the books by Chatfield and by Velleman and Hoaglin listed in the bibliography at the end of this chapter.

6.3   THE SIGN TEST

The **sign test** is amongst the simplest of all non-parametric methods, and was first discussed in the early eighteenth century. It can be used in a number of ways, the simplest of which is demonstrated by the following example.

> *Example.*   A pharmaceutical preparation is claimed to have a median content of 8% of a particular component. Successive batches were found in practice to contain 7.3, 7.1, 7.9, 9.1, 8.0, 7.1, 6.8 and 7.3% of the constituent. At the 5% significance level, does this indicate that the claimed median is incorrect?

In Chapter 3 (Section 3.2) it was shown that such problems could be tackled by using the t-test after calculation of the mean and standard deviation of the experimental data. The t-test assumes, however, that the data are normally distributed. The sign test avoids such an assumption, and is much easier to perform. The postulated median is subtracted from each experimental value in turn, and the **sign** of each result is considered. Values equal to the postulated median are ignored *entirely*. In this case, therefore, we effectively have seven experimental values, six of them lower than the median and hence giving minus signs, and one higher than the median and hence giving a plus sign. To test whether this preponderance of minus signs is significant we use the binomial theorem. This theorem shows that the probability of r out of n signs being minus is given by

$$P(r) = {}^nC_r p^r q^{(n-r)} \tag{6.1}$$

where ${}^nC_r$ is the number of combinations of r items from a total of n items, p is the probability of getting a minus sign in a single result, and q is the probability of not getting a minus sign in a single result, i.e. $q = 1 - p$. Since the median is defined so that half the experimental results lie above it, and half below it, it is clear that if the median *is* 8.0 in this case, then both p and q should be $\frac{1}{2}$. Using Eq. (6.1), we find that $P(6) = {}^7C_6 \times (\frac{1}{2})^6 \times \frac{1}{2} = 7/128$. Similarly we can calculate that the chance of getting 7 minus signs, $P(7)$, is 1/128. Overall, therefore, the probability of getting 6 *or more* negative signs in our experiment is 8/128. We are only asking, however, whether the data differ significantly from the postulated median. We must therefore perform a two-tailed test (cf. Chapter 3), i.e. we must calculate the probability of obtaining 6 or more identical signs (i.e. $\geqslant 6$ plus or $\geqslant 6$ minus signs) when 7 results are taken at random. This is clearly 16/128 = 0.125. We then compare this figure with 0.05, i.e. we perform our test at the 5% significance level. Since the experimental value is > 0.05, our null hypothesis, i.e. that the data come from a population of median 8.0, cannot be rejected. As in Chapter 3, it is important to note that we have not proved that the data *do* come from such a population; we have only concluded that such a hypothesis cannot be rejected.

It is apparent from this example that the sign test will involve the frequent use of the binomial distribution with $p = q = \frac{1}{2}$. So common is this approach to non-parametric statistics that most sets of statistical tables include the necessary data, allowing such calculations to be made instantly (see Table A.7). Moreover, in many practical situations, an analyst will always use the same value of n, i.e. he will always take the same number of readings or samples, and he will be able to memorize easily the probabilities corresponding to the various numbers of $+$ or $-$ signs.

The sign test can also be used as a non-parametric alternative to the paired t-test (Section 3.4) to compare two sets of results for the same samples. Thus if ten samples are examined by each of two methods, A and B, we can test whether the two methods give significantly different readings, by calculating for each sample the difference in the results, i.e. [(result obtained by method A) $-$ (result obtained by method B)]. The null hypothesis will be that the two methods do not give significantly different results — in practice this will again mean that the probability of obtaining a plus sign (or a minus sign) is 0.5. The number of plus or minus signs actually obtained can be compared with the probability derived from Eq. (6.1). An example of this application of the sign test is given in the exercises at the end of the chapter.

A further use of the sign test is to indicate a trend. This application is illustrated by the following example.

Example. The level of a hormone in a patient's blood plasma is measured at the same time each day for 10 days. The resulting data are:

Day	1	2	3	4	5	6	7	8	9	10
Level, ng/ml	5.8	7.3	4.9	6.1	5.5	5.5	6.0	4.9	6.0	5.0

Is there any evidence for a trend in the hormone concentration?

Using parametric methods, it would be possible to make a linear regression plot of such data and test whether its slope differed significantly from zero (Chapter 5). Such an approach would assume that the errors were normally distributed, and that any trend that did occur was linear. The non-parametric approach is again simpler. The data are divided into two equal sets, the sequence being retained:

5.8 7.3 4.9 6.1 5.5
5.5 6.0 4.9 6.0 5.0

The result for the sixth day is then subtracted from that for the first day, that for the seventh day from that for the 2nd day etc. The signs of the differences between the pairs of values in the five columns are determined in this way to be $+, +, 0, +, +$. As usual the zero is ignored, leaving 4 results, all positive. The probability of obtaining four identical signs in four trials is clearly $2 \times (1/16) = 0.125$. (Note that a two-tailed test is again used, as the trend in the hormone level might be upwards or downwards.) The null hypothesis, that there is no trend in the results, can thus not be rejected at the 5% significance level.

The critical reader may find this result rather unsatisfactory. On inspection of the data it is hard to avoid the conclusion that, despite the result of the sign test, there is in fact a downward trend in the results. For example, the mean of the first three results is 6.00, the mean of the next four is 5.78, and the mean of the last three is 5.30. This apparent discrepancy is partly a result of the nature of hypothesis testing: we cannot reject the hypothesis that there is no trend in the results, but that does not mean that such a trend *must necessarily* be absent. The example is also a reminder that the price paid for the extreme simplicity of the sign test is some loss of statistical power. The test does not utilize all the information offered by the data, so it is not surprising to find that it also provides less discriminating information. In later sections, non-parametric methods that *do* use the magnitudes of the individual results as well as their signs will be discussed.

6.4 THE WALD–WOLFOWITZ RUNS TEST

In some instances we are interested not merely in whether observations generate positive or negative signs, but also in whether these signs occur in a random sequence. In Section 5.11, for example, we showed that if a straight line is a good fit to a set of calibration points, positive and negative residuals will occur more or less at random. By contrast, attempting to fit a straight line to a set of points that actually lie on a curve will yield non-random sequences of positive or negative signs: there might, for example, be a sequence of $+$ signs, followed by a sequence of $-$ signs, and then another sequence of $+$ signs. Such sequences are technically known as **runs** — the word being used here in much the same way as when someone refers to 'a run of bad luck', or when a sportsman suffers 'a run of low scores'. In the curve-fitting case, it is clear that a non-random sequence of $+$ and $-$ signs will lead to a smaller number of runs than a random sequence. The Wald–Wolfowitz method tests whether the number of runs is small enough for the null hypothesis of a random distribution of signs to be rejected. The number of runs in the experimental data is compared with the numbers in Table A.8, which refers to the 5% significance level. The table is entered by using the appropriate values for N, the number of $+$ signs, and M, the number of $-$ signs. If the experimental number of runs is *smaller* than the tabulated value, then the null hypothesis can be rejected.

Example.   Linear regression equations are used to fit a straight line to a set of 12 calibration points. The signs of the resulting residuals in order of increasing x value are: $+ + + + - - - - - - + +$. Comment on whether it would be better to attempt to fit a curve to the points.

Here it is clear that $N = M = 6$, and that the number of runs is 3. Table A.8 shows that, at the 5% significance level, the number of runs must be < 4 if the null hypothesis is to be rejected. Thus in this instance we can reject the null hypothesis, and conclude that the sequence of $+$ and $-$ signs is not a random one. The attempt to fit a straight line to the experimental points is therefore unsatisfactory, and a curvilinear regression plot is indicated instead.

There are two further points of interest in connection with the Wald–Wolfowitz test. First, it is worth noting that it can be used with any results that can be divided or converted into just two categories. Suppose, for example, that it is found that twelve successively used spectrometer light-sources last for 450, 420, 500, 405, 390, 370, 380, 395, 370, 370, 420, and 430 hr. The median lifetime, in this case the average of the sixth and seventh numbers when the data are arranged in ascending order, is 400 hr. If all those lamps with lifetimes less than the median are given a $-$ sign, and those with longer lifetimes are given a plus sign, then the sequence becomes: $+$ $+$ $+$ $+$ $-$ $-$ $-$ $-$ $-$ $-$ $+$ $+$. This is the same sequence as in the regression example above, and is shown in the same way to be significantly non-random. In this case, the significant variations in lifetime might be explained if the lamps came from different batches or different manufacturers.

It is also important to note that we may be concerned with unusually large numbers of short runs, as well as unusually small numbers of long runs. Thus, if 6 $+$ and 6 $-$ signs occurred in the order $+$ $-$ $+$ $-$ $+$ $-$ $+$ $-$ $+$ $-$ $+$ $-$, we would strongly suspect a non-random sequence. Table A.8 shows that, with $N = M = 6$, a total of 11 or 12 runs indicates that the null hypothesis of random order should be rejected, and some periodicity in the data suspected.

6.5   THE WILCOXON SIGNED RANK TEST

Section 6.3 described the use of the sign test. Its value lies in the minimal assumptions it makes about the experimental data. The population from which the sample is taken is not assumed to be normal, or even to be symmetrical — the only prior information needed is the value of the median. A compensating disadvantage of the sign test is that it uses so little of the information provided. The only material point is whether an individual measurement is greater than or less than the median — the magnitude of this deviation is not used at all.

In many instances an analyst will have every reason to believe that his measurements will be *symmetrically* distributed but will not wish to make the assumption that they are normally distributed. This assumption of symmetrical data, and the consequence that the mean and the median of the population will be equal, allows more powerful significance tests to be developed. Important advances were made by Wilcoxon, and his signed rank test has several applications. Its mechanism is best illustrated by an example.

Example.   The blood lead levels (in pg/ml) of seven children were found to be 104, 79, 98, 150, 87, 136 and 101. Could such data come from a population, assumed to be symmetrical, with median (mean) of 95 pg/ml?

Compared with the reference value (95) the data have values of

$$9, -16, 3, 55, -8, 41, 6$$

These values are first arranged in order of magnitude without regard to sign:

$$3, 6, 8, 9, 16, 41, 55$$

Their signs are then restored to them (in practice these last two steps would of course be combined):

$$3, 6, -8, 9, -16, 41, 55$$

The numbers are then **ranked**; in this process they keep their signs but are assigned numbers indicating their order (or rank):

$$1, 2, -3, 4, -5, 6, 7$$

The positive ranks add up to 20, and the negative ones to 8. The *lower* of these two figures (8) is taken as the test statistic. The binomial theorem will give the probability of this number occurring. If the data come from a population with median 95 the sums of the negative and positive ranks would be expected to be approximately equal numerically; if the population median was very different from 95 the sums of the negative and positive ranks would be unequal. The probability of a particular sum occurring in practice is given by a set of tables (see Table A.9). In this test the null hypothesis is rejected if the experimental value is *less than or equal to* the tabulated value, i.e. the opposite of the situation encountered in most significance tests. In the present example, examination of Table A.9 shows that, for $n = 7$, the test statistic must be less than or equal to 2 before the null hypothesis — that the data *do* come from a population of median (mean) 95 — can be rejected at a significance level of $P = 0.05$. In this example, the null hypothesis must clearly be retained. As usual, a two-tailed test is used, though there may be occasional cases where a one-tailed test is more appropriate.

An important advantage of the signed rank test is that it can also be used on paired data, because they can be transformed into the type of data given in the previous example. The signed rank method can thus be used as a non-parametric alternative to the paired t-test (Section 3.4).

Example. The following table gives the percentage concentration of zinc, determined by two different methods, for each of eight samples of health food.

Sample	EDTA titration	Atomic spectrometry
1	7.2	7.6
2	6.1	6.8
3	5.2	4.6
4	5.9	5.7
5	9.0	9.7
6	8.5	8.7
7	6.6	7.0
8	4.4	4.7

Is there any evidence for a systematic difference between the results of the two methods?

The approach to this type of problem is very simple. If there were no systematic difference between the two methods, then we would expect that the differences between the results for each sample, i.e. [(titration result) − (spectrometry result)], should be symmetrically distributed about zero. The signed differences are easily shown to be:

$$-0.4, \; -0.7, 0.6, 0.2, \; -0.7, \; -0.2, \; -0.4, \; -0.3$$

Arranging these values in numerical order without regard to sign, we have:

$$-0.2, 0.2, \; -0.3, \; -0.4, \; -0.4, 0.6, \; -0.7, \; -0.7$$

It is clear that the ranking of these results presents a difficulty, that of *tied ranks*. There are two results with the numerical value 0.2, two with a numerical value of 0.4, and two with a numerical value of 0.7: how are the ranks to be calculated? This problem, in practice the only one encountered in ranking methods, is resolved by giving the tied values average ranks, with appropriate signs. Thus the ranking for the present data is:

$$-1.5, 1.5, \; -3, \; -4.5, \; -4.5, 6, \; -7.5, \; -7.5$$

In such cases, it is worth verifying that the ranking has been done correctly by calculating the sum of all the ranks without regard to sign. The sum for the numbers above is 36, which is the same as the sum of the first eight natural numbers, and therefore correct. The sum of the positive ranks is 7.5, and the sum of the negative ranks is 28.5. The test statistic is thus 7.5. Inspection of Table A.9 shows that, for $n = 8$, the test statistic has to be ≤ 3 before the null hypothesis can be rejected at the significance level $P = 0.05$. In the present case, the null hypothesis must be retained — there is no evidence that the median (mean) of the difference is not zero, and hence no evidence for a systematic difference between the two analytical methods.

The signed rank test is seen from these examples to be a simple and valuable method. Its principal limitation is that it cannot be applied to very small sets of data: for a two-tailed test at the significance level $P = 0.05$, n must be at least 6.

6.6   THE WILCOXON RANK SUM TEST AND RELATED METHODS

The signed rank test described in the previous section is valuable for the study of single sets of measurements, and for paired sets that can readily be reduced to single sets. In many instances, however, it is necessary to compare two independent samples that cannot be reduced to a single set of data. Such samples may contain different numbers of measurements. Several non-parametric tests to tackle such problems have been devised. The simplest to understand (though not, perhaps, the simplest to perform) is the Wilcoxon rank sum test. The operation of this test is most easily demonstrated by an example.

Example.   A sample of photographic waste was analysed for silver by atomic-absorption spectrometry, five successive measurements giving values of 9.8, 10.2, 10.7, 9.5, and 10.5 μg/ml. After chemical treatment, the waste was analysed again by the same procedure, five successive measurements giving values of 7.7, 9.7, 8.0, 9.9 and 9.0 μg/ml. Is there any evidence that the treatment produced a significant reduction in the levels of silver?

The first step in the calculation is to rank *all* the data (i.e. from both sets of measurements). To distinguish the results obtained after the treatment we underline them:

$$\underline{7.7}, \underline{8.0}, \underline{9.0}, 9.5, \underline{9.7}, 9.8, \underline{9.9}, 10.2, 10.5, 10.7$$

The rankings, with those for the treated sample measurements again underlined, are:

$$\underline{1}, \underline{2}, \underline{3}, 4, \underline{5}, 6, \underline{7}, 8, 9, 10$$

It is apparent that, even when tied ranks arise and are treated as shown in the previous section, the sum of all the ranks must be $n(n + 1)/2$, in this case 55.

If the measurements for the two sets of results were indistinguishable, we would expect that the underlined ranks and the other ranks should be mixed more or less at random in the list. That is, the sum of the underlined ranks and the sum of the other ranks should be similar (if the numbers of measurements are the same in the two sets of results). In the present case the sum of the underlined ranks (treated samples) is 18 and the sum of the other ranks (untreated samples) is 37. These rank sums must now be converted into the test statistics T_1 and T_2, by using the equations:

$$T_1 = S_1 - n_1(n_1 + 1)/2 \tag{6.5}$$

$$T_2 = S_2 - n_2(n_2 + 1)/2 \tag{6.6}$$

where S_1 is the sum of one set of ranks and S_2 the sum of the other (18 and 37 in this example) and n_1 and n_2 are the corresponding numbers of observations. In this case $n_1 = n_2 = 5$, so $n_i(n_i + 1)/2 = 15$, and T_1 and T_2 are 3 and 22 respectively. The lower of these two values (3) is compared with the appropriate value in Table A.10. It is particularly important to note that in this example we can legitimately apply a *one-tailed* test. This is because the treatment method is specifically designed to *reduce* the silver level of the photographic waste: it may or may not do this successfully, but there is no realistic chance that it will actually increase the silver level! The table shows that, for $P = 0.05$ and $n_1 = n_2 = 5$, the lower T value must not exceed 4 if the null hypothesis is to be rejected. It is thus possible to reject the null hypothesis — that the two samples came from the same population — and to conclude that the chemical treatment was effective in reducing the silver level in the photographic waste.

The operation of the rank sum test is complicated by the calculation of the T values from the rank sums. This additional step is essential, however, because the test is also valid when n_1 and n_2 are not equal. In such instances it is not reasonable to expect the sums of the two sets of ranks to be equal even if they are mixed more or less at random in the overall ranking, and Eqs. (6.5) and (6.6) make allowance for this effect. Because of this additional calculation two alternative, and probably simpler, tests have been devised to compare two sets of unpaired data.

The first, the **Mann-Whitney U-test**, is just an easier but equally valid method of calculating the lower T value defined above.

Example.   Apply the Mann-Whitney U-test to the data of the previous example.
The procedure involves finding the number of results in one sample that exceed each of the values in the other sample. In the present case, we believe that the silver concentration of the treated solution should, if anything, be lower than that of the untreated solution. We thus expect to find that the number of cases in which a treated sample has a higher value than an untreated one should be small. Each of the values for the untreated sample is listed, and the number of instances where the values for the treated sample are greater are noted in each case.

Untreated sample	Higher values in treated sample	Number of higher values
9.8	9.9	1
10.2	—	0
10.7	—	0
9.5	9.7, 9.9	2
10.5	—	0

The total of the numbers in the third column is 3, the same T value as that obtained in the rank sum test. It may be shown that the two methods always produce the same result, so Table A.10 is used for the Mann-Whitney U-test as well as for the rank sum test. It is clear that the U-test method is a much simpler way of arriving at the T value. Indeed, when the numbers of measurements are small the calculation can be done mentally, a great advantage. If ties (identical values) occur in the U-test, each tie is assigned a value of 0.5 in the count of T.

A further convenient method with some interesting features is **Tukey's quick test**. It is appropriate to exemplify its use with the same example.

Example. Apply Tukey's quick test to the data of the previous example.

The test can be regarded as having two stages, though when only a few results are available, these two steps will doubtless be amalgamated into one rapid mental calculation. In the first step, the number of results in the second set of data that are *lower than all the values in the first set* are counted. If there are *no* such values, the test ends at once, and the null hypothesis of equal medians is accepted. In the present example, there are three such values, the readings 7.7, 8.0 and 9.0 being lower than the lowest value from the first set (9.5). The test thus continues to the second step, in which we count all the values in the first data set that *are higher than all the values in the second set*. Again, if there are no such values, the test ends and the null hypothesis is accepted. Here, there are again three such values, the readings 10.2, 10.5 and 10.7 exceeding the highest value in the second set (9.7). (This approach may be contrasted with that of the Mann-Whitney U-test, which identifies *high* values in the sample which might be expected to have the *lower* median.) Overall, therefore, there are six values that are not within the range over which the two samples overlap. This total (often called T, but not to be confused with the T statistics in the rank sum test) is the test statistic. The most interesting and valuable aspect of Tukey's quick test is that statistical tables are not normally needed to interpret this result. Provided that the number of readings in each sample does not exceed about 20, and that the two sample sizes are not greatly different (conditions that would probably be valid in most analytical laboratory experiments), the reference values of T for a particular level of significance are *independent of sample size*. For a one-tailed test the null hypothesis may be rejected if $T \geq 6$ (*for* $P = 0.05$), ≥ 7 ($P = 0.025$), ≥ 10 ($P = 0.005$), and ≥ 14 ($P = 0.0005$). (For a two-tailed test the critical T values at $P = 0.05, 0.025, 0.005$ and 0.0005 are 7, 8, 11 and 15 respectively). In the present example, therefore, the experimental T value is big enough to be significant at $P = 0.05$ in a one-tailed test. We can thus reject the null hypothesis and report that the treatment does reduce the silver content of the photographic waste significantly, a result in accord with that of the Mann–Whitney U-test.

If ties occur in Tukey's quick test (i.e. if one of the values in the hypothetically higher sample is equal to the highest value in the other sample, or if one of the values in the 'lower' sample is equal to the lowest value in the 'higher' sample) then each tie counts 0.5 towards the value of T.

Apart from its normal application to the comparison of medians, the Wilcoxon rank sum test has been modified by Siegel and Tukey to compare the spread of two sets of results, and thus offer a genuinely non-parametric alternative to the F-test (cf. Section 3.5). In this version of the rank sum test, the data are pooled and arranged in numerical order, as before, with one set of results distinguished by underlining. Then, however, they are ranked in a different way: the lowest measurement is ranked one, the highest measurement ranked two, the highest but one measurement ranked three, the lowest but one ranked four, the lowest but two ranked five, and so on. (If the total number of measurements is odd, the central measurement is ignored). This 'paired alternate ranking' produces a situation in which the low and high results receive low ranks, and the central results receive high ranks. If one data set has a significantly wider spread than the other, its sum of ranks should thus be much lower, while if the dispersion of the two sets of results is similar, their rank sums will be similar. Application of this method to the example used earlier in this section gives the following rankings:

Data:	7.7,	8.0,	9.0,	9.5,	9.7,	9.8,	9.9,	10.2,	10.5,	10.7
Ranks:	1	4	5	8	9	10	7	6	3	2

The sum of the underlined ranks (treated silver-containing samples) is 26, and the rank sum for the untreated samples is 29. These sums are then converted into the test statistics T_1 and T_2 as in Eqs. (6.5) and (6.6). The lower of these values, the one used in the test, is thus 11, much higher than the critical value of 2 (for a *two*-tailed test, at $P = 0.05$) (Table A.10). The null hypothesis, in this case that the spread of the results is similar for the two sets of data, is thus retained. A little thought will show that the validity of this useful test will be reduced if the average values for the two sets of data are substantially different. In the extreme case where all the measurements in one sample are lower than all the measurements in the other sample, the values of T_1 and T_2 will always be as similar as possible, whatever the spread of the two samples. If this effect is feared to be appreciable, it is permissible to estimate the means of the two samples, and add the difference between the means to each of the measurements of the lower set. This will remove any effect due to the different means, while preserving the dispersion of the sample. An example of the application of this test is provided at the end of the chapter.

6.7 NON-PARAMETRIC TESTS ON MORE THAN TWO SAMPLES

The previous section described tests in which two statistical samples were compared with each other. Non-parametric methods are not, however, limited to two sets of data: several methods which compare three or more samples are available. Before two of these tests are outlined, it is important to mention one pitfall to be avoided in all multi-sample comparisons. When (for example) three sets of measurements are examined to see whether or not their medians are similar, there is a great temptation to compare only the two samples with the highest and lowest medians. This simplistic approach can give misleading results. When several samples are taken from the *same parent population*, there are cases where the highest and lowest medians, considered in isolation, appear to be significantly different. This is because, as the number of samples increases, the difference between the highest and the lowest medians will tend to increase. The correct approach is to perform first a test that considers *all the samples together*: if it shows that they might not all come from the same population, *then* separate tests may be performed to try to identify where the significant differences occur. Here we describe in outline the principles of two non-parametric tests for three or more sets of data: the reader seeking further detail is recommended to consult the books given in the bibliography.

The **Kruskal–Wallis test** may be regarded as an extension of the Wilcoxon rank sum procedure to the comparison of the medians of three or more unmatched samples. (An extension of the silver analysis described in the previous section, with three samples of photographic waste, one untreated and the other two treated by different methods, would provide an instance where the test would be useful). The results from the three (or more) samples are pooled and arranged in rank order. The rank totals for the data from the different samples are determined: tied ranks are averaged, as in the Wilcoxon method, though a special correction procedure is advisable if there are numerous ties. It is easy to appreciate that, if each sample has the same number of measurements (this is not a requirement of the test), and if the samples have similar medians, then the rank totals for each sample should be similar, so the sum of their squares should be a minimum. Thus, if we have three samples, each with five measurements, the rankings will range from 1 to 15 and the sum of all the ranks will be 120. Suppose that the three medians are very similar, and that the rank totals for each sample are thus equal, each being 40. The sum of the squares of these totals will thus be $40^2 + 40^2 + 40^2 = 4800$. If the medians are significantly different, then the rank totals will also be different from one another — say 20, 40, and 60. The sum of the squares of such totals will always be larger than 4800 ($20^2 + 40^2 + 60^2 = 5600$).

The probability of obtaining any particular sum of squares can be determined by using the **chi-squared** statistic (cf. Chapter 3). If the samples are referred to as A, B, C etc. (k samples in all), with numbers of measurements n_A, n_B, n_C etc. and rank totals R_A, R_B, R_C etc., then the value of χ^2 is given by:

$$\chi^2 = \frac{12}{N^2 + N} \left(\frac{R_A^2}{n_A} + \frac{R_B^2}{n_B} + \frac{R_C^2}{n_C} + \ldots \right) - 3(N+1) \tag{6.7}$$

where $N = n_A + n_B + n_C$ etc. This χ^2 value is compared as usual with tabulated χ^2 values. The latter are identical to the usual values when the total number of measurements is > ca. 15, but special tables are used for smaller numbers of measurements. The number of degrees of freedom is $k - 1$. Experimental values of χ^2 that exceed the tabulated values allow the null hypothesis (that the medians of the samples are not significantly different) to be rejected. As already noted, in the latter situation further tests can be performed on individual pairs of samples: again, the reader is referred to the bibliography for more details.

We have already seen (Sections 3.4 and 6.3) that when *paired* results are compared, special statistical tests can be used. These tests use the principle that, when two experimental methods which do not differ significantly are applied to the same chemical samples, the differences between the matched pairs of results should be close to zero. This principle can be extended to three or more matched sets of results by using a non-parametric test devised in 1937 by the well-known American economist, Milton Friedman. In analytical chemistry, the main application of **Friedman's test** is in the comparison of three (or more) experimental methods applied to the same chemical samples. The test again uses the χ^2 statistic, in this case to assess the differences that occur between the total rank values for the different methods. The following example illustrates the simplicity of the approach:

Example.   The levels of a pesticide in 4 plant extracts were determined by (A) high-performance liquid chromatography, (B) gas–liquid chromatography, and (C) radioimmunoassay. The following results (all in ng/ml) were obtained:

Sample	Method		
	A	B	C
1	4.7	5.8	5.7
2	7.7	7.7	8.5
3	9.0	9.9	9.5
4	2.3	2.0	2.9

Do the three methods give values for the pesticide levels that differ significantly?

This problem is simply solved by replacing the numbers in the results table by ranks. For each sample the method giving the lowest result is ranked 1, and that giving the highest result is ranked 3, as follows:

Sample	Method		
	A	B	C
1	1	3	2
2	1.5	1.5	3
3	1	3	2
4	2	1	3

The use of an average value is necessary in the case of tied ranks in sample 2 (cf. Section 6.5). The sums of the ranks for the three methods A, B and C are 5.5, 8.5 and 10 respectively. These sums should total $nk(k + 1)/2$, where k is the number of methods (here three) and n the number of samples (here four). The rank sums are squared, yielding 30.25, 72.25 and 100 respectively, and these squares are added to give the statistic R, which in this case is 202.5. The experimental value of χ^2 is then calculated from:

$$\chi^2 = \frac{12R}{nk(k + 1)} - 3n(k + 1) \tag{6.8}$$

which gives a result of 2.625. At the significance level $P = 0.05$, and with $k = 3$, the critical values of χ^2 are 6.0, 6.5, 6.4, 7.0, 7.1 and 6.2 for $n = 3, 4, 5, 6, 7$ and 8 respectively. (More extensive data are given in many sets of statistical tables, and when $k > 7$, the usual χ^2 tables can be used at $k - 1$ degrees of freedom). In this instance, the experimental value of χ^2 is much less than the critical value, and we must clearly retain the null hypothesis: the three methods give results that do not differ significantly.

It is noteworthy that the Friedman test could alternatively be used in the reverse form: assuming that the three analytical methods give indistinguishable results, the same procedure could be used to test differences between the four plant extracts. In this case k and n are 4 and 3 respectively, and the reader may care to verify that R is 270 and that the resulting χ^2 value is 9.0. This is higher than the critical value for $P = 0.05$, $n = 3$, $k = 4$, which is 7.4. So in this second application of the test we can reject the null hypothesis, and state that the four samples _do_ differ in their pesticide levels. Further tests, which would allow selected comparisons between pairs of samples, are then available.

Friedman's test is evidently much simpler to perform in practice than the ANOVA method (Sections 3.8–3.10), though it does not have the latter's ability to study interaction effects (see Chapter 7).

6.8 RANK CORRELATION

Ranking methods can also be applied to correlation problems. The Spearman rank correlation coefficient method to be described in this section is the oldest application of ranking methods in statistics, dating from 1904. Like other ranking methods, it is of particular advantage when one or both of the sets of observations under investigation can be expressed *only* in terms of a rank order rather than in quantitative units. Thus, in the following example, the possible correlation between the sulphur dioxide concentrations in a series of table wines and their taste quality is investigated. The taste quality of a wine is not easily expressed in quantitative terms, but it is relatively simple for a panel of wine-tasters to rank the wines in order of preference. Examples of other attributes that are easily ranked, but not easily quantified, include the condition of experimental animals, the quality of laboratory accommodation, and the efficiency of laboratory staff. It should also be remembered that if either or both the sets of data under study should happen to be quantitative, then (in contrast to the methods described in Chapter 5) there is no need for them to be normally distributed. Like other non-parametric statistics, the Spearman rank correlation coefficient, ρ, is easy to determine and interpret. This is shown in the following example.

Example. Seven different table wines are ranked in order of preference by a panel of experts; the best wine is ranked 1, the next best 2, and so on. The sulphur dioxide content (in parts per million) of each wine is then determined by flow-injection analysis with colorimetric detection. Use the following results to determine whether there is any relationship between perceived wine quality and sulphur dioxide content.

Wine:	A	B	C	D	E	F	G
Taste ranking:	1	2	3	4	5	6	7
SO$_2$ content:	0.9	2.8	1.7	2.9	3.5	3.3	4.7

The first step in the calculation is to convert the sulphur dioxide concentrations from absolute values into rank values (if tied ranks occur they are averaged as described in previous sections):

Wine:	A	B	C	D	E	F	G
Taste ranking:	1	2	3	4	5	6	7
SO$_2$ content:	1	3	2	4	6	5	7

The differences, d_i, between the two ranks are then calculated. They are clearly 0, -1, 1, 0, -1, 1, 0. The correlation coefficient, ρ, is then given by:

$$\rho = 1 - \frac{6\sum_i d_i^2}{n(n^2 - 1)} \tag{6.9}$$

In this example, ρ is $1 - (24/336)$, i.e. 0.929. Theory shows that, like the product-moment correlation coefficient, ρ can vary between -1 and $+1$. When $n = 7$, ρ must exceed 0.786 if the null hypothesis of no correlation is to be rejected at the significance level $P = 0.05$ (Table A.11). Here, we can conclude that there is a correlation between the sulphur dioxide content of the wines and their perceived quality. Bearing in mind the way the rankings were defined, there is strong evidence that higher sulphur dioxide levels produce less palatable wines!

Another rank correlation method, due to Kendall, was introduced in 1938. It claims to have some theoretical advantages over the Spearman method, but is harder to calculate (especially when tied ranks occur) and is not so frequently used.

6.9 NON-PARAMETRIC REGRESSION METHODS

In the detailed discussion of linear regression methods in the previous chapter, the assumption of normally distributed y-direction errors was emphasized, and the complexity of some of the calculation methods was apparent. This complexity is largely overcome by using calculators or computers, and there are also some rapid approximate methods for fitting straight lines to experimental data (see bibliography). There remains an interest in non-parametric approaches to fitting a straight line to a set of points: of the several methods available, perhaps the simplest is **Theil's 'incomplete' method**, so called to distinguish it from another more complex procedure developed by the same author (the 'complete' method).

Theil's method assumes that a series of points (x_1, y_1), (x_2, y_2), etc. is fitted by a line of the form $y = bx + a$. The first step in the calculation involves ranking the points in order of increasing x. If the number of points, x, is odd, the middle point, i.e. the median value of x, is deleted: the calculation always requires an even number of points. For any pair of points (x_i, y_i), (x_j, y_j) where $x_j > x_i$, the slope, b_{ij}, of the line joining the points can be calculated from:

$$b_{ij} = (y_j - y_i)/(x_j - x_i) \tag{6.10}$$

Slopes b_{ij} are calculated for the pair of points (x_1, y_1) and the point immediately after the median x-value, for (x_2, y_2) and the second point after the median x-value, and so on until the slope is calculated for the line joining the point immediately before the median x with the last point. Thus, if the original data contained 11 points, 5 slopes would be estimated (the median point having been omitted); 8 original points would give 4 slope estimates, and so on. These slope estimates are themselves arranged in ascending order and their median value, calculated as described in Section 6.2, is the estimated slope of the straight line. With this value of b, values a_i for the intercept are estimated for each point with the aid of the equation $y = bx + a$. Again the estimates of a are arranged in ascending order and the median value is chosen as the best estimate of the intercept of the line. The method is illustrated by the following example.

Example. The following results were obtained in a calibration experiment for the absorptiometric determination of a metal chelate complex:

Concentration, μg/ml:	0	10	20	30	40	50	60	70
Absorbance:	0.04	0.23	0.39	0.59	0.84	0.86	1.24	1.42

Use Theil's method to estimate the slope and the intercept of the best straight line through these points.

In this case the calculation is simplified by the occurrence of an even number of observations, and by the fact that the *x*-values (i.e. the concentrations) occur at regular intervals and are already in ranking order. We thus calculate slope estimates from 4 pairs of points:

$$b_{15} = (0.84 - 0.04)/40 = 0.0200$$
$$b_{26} = (0.86 - 0.23)/40 = 0.0158$$
$$b_{37} = (1.24 - 0.39)/40 = 0.0212$$
$$b_{48} = (1.42 - 0.59)/40 = 0.0208$$

We now arrange these slope estimates in order, obtaining 0.0158, 0.0200, 0.0208, 0.0212. The median estimate of the slope is thus the average of 0.0200 and 0.0208, i.e. 0.0204. We now use this value of *b* to estimate the intercept, *a*; the eight individual a_i values are:

$$a_1 = 0.04 - (0.0204 \times 0) = +0.040$$
$$a_2 = 0.23 - (0.0204 \times 10) = +0.026$$
$$a_3 = 0.39 - (0.0204 \times 20) = -0.018$$
$$a_4 = 0.59 - (0.0204 \times 30) = -0.022$$
$$a_5 = 0.84 - (0.0204 \times 40) = +0.024$$
$$a_6 = 0.86 - (0.0204 \times 50) = -0.160$$
$$a_7 = 1.24 - (0.0204 \times 60) = +0.016$$
$$a_8 = 1.42 - (0.0204 \times 70) = +0.008$$

Arranging these intercept estimates in order, we have -0.160, -0.022, -0.018, -0.008, $+0.016$, $+0.024$, $+0.026$, $+0.040$. The median estimate is $+0.004$. We have thus concluded that the best straight line is given by $y = 0.0204x + 0.004$. The 'least-squares' line, calculated by the methods of Chapter 5, is readily shown to be $y = 0.0195x + 0.019$. Figure 6.1 shows that the two lines are very similar when plotted. However, the Theil method has three distinct advantages: it does not assume that all the errors are in the *y*-direction; it does not assume that either the *x*- or *y*-direction errors are normally distributed; and it is not affected by the presence of outlying results. This last point is clearly illustrated by the point (50,0.86) in the present example. It has every appearance of being an outlier, but its value does not affect the Theil calculation at all, since neither b_{26} nor a_6 directly affects the median estimates of the slope and intercept respectively. In the least-squares calculation, however, this outlying point carries as much weight as the other points. This is reflected in the calculated results; the least-squares line passes closer to the outlier than the non-parametric line does.

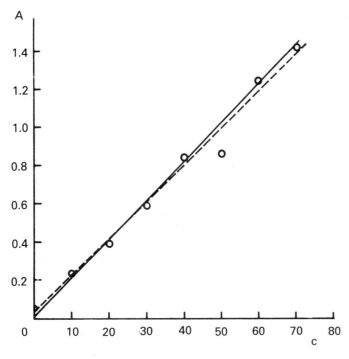

Fig. 6.1.   Straight-line calibration graph calculated by Theil's method (———), and by the
least-squares method of Chapter 5 (———).

Statistical methods that are not greatly affected by outlying results are said to be
'robust' methods; their value in this and other cases is clear.

Unlike most non-parametric methods, Theil's method involves tedious calculations, and a computer program would be a great advantage. It should be noted that
non-parametric methods for fitting curves are also available, but these are beyond
the scope of the present book.

6.10 ROBUST METHODS

At the beginning of this chapter, it was noted that there is growing evidence for
the occurrence in the experimental sciences of *heavy-tailed* error distributions, which
can be regarded as arising from normal (Gaussian) distributions with the addition
of outliers arising from gross errors, or from the superposition of several normal
distributions with similar means but different variances. In either case, it seems to
be a waste of information to use non-parametric methods, which make no assumptions at all about the underlying error distribution. A better approach would be to
develop methods which do not exclude entirely data which are suspiciously different
from the rest of the measurements or from the mean value, but which *reduce the
weight* given to such data. This is the philosophy underlying the robust methods to
be summarized in this section. Such methods can be applied to repeated measurements and also to calibration/regression data. Just as in other areas of statistics, a

very large number of robust methods has been developed, so only a brief survey of this rapidly developing field is possible here: the reader is referred to the bibliography for sources of further material. An obvious problem occurs in virtually all these methods. If we are to downgrade the significance of some of our measurements, one or more criteria are needed on which to base such decisions, but we cannot use such criteria unless we initially consider all the data. This conundrum is circumvented by using *iterative* methods: we estimate or guess a starting value or values for some property of our data, use such initial estimates with our weighting criteria to arrive at a second estimate, then we re-apply our criteria etc. Such methods are only possible without great inconvenience if a computer is available, though it must be stressed that many otherwise excellent suites of statistics software do not (yet) include programs for robust methods.

There are some very simple robust methods which do not require such iterations, because they arbitrarily eliminate, rather than downweight, a proportion of the data. For example, it is easy to calculate a *trimmed mean* for any set of data by omitting the smallest r and the largest r observations. This principle can be applied to the set of data in one of the examples in Section 3.7. This example considered seven replicate measurements of nitrite ion in river water (mg/l):

0.380, 0.400, 0.401, 0.403, 0.410, 0.411, 0.413

The data have been arranged in numerical order for convenience: this emphasizes that the obvious question is whether or not the measurement 0.380 is an outlier. First we note that, if the number 0.380 is retained, the mean of the seven measurements is 0.4026, and their standard deviation is 0.0112. If, as the calculation on p. 64 suggests is permissible at $P = 0.05$, we reject the result 0.380, the mean and standard deviation become 0.4063 and 0.0056 respectively. This confirms, as noted in Section 3.3, that the mean and (especially) the standard deviation are vulnerable to the occurrence of outliers. Now suppose that we omit the smallest (0.380) and the largest (0.413) of the above measurements, and recalculate the mean. This produces a number technically known as the 14.28% trimmed mean, the percentage being calculated as $100r/n$ where r top and bottom measurements have been omitted from n results. The result is 0.4050, clearly closer to the second of the two means calculated above, i.e. the mean determined after rejection of the outlier. The robustness of this trimmed mean is obvious — it would have been the same, *whatever* the values taken by the smallest and largest results. But this also illustrates the crudity of the trimmed mean method. Why should we omit the value 0.413, except for reasons of symmetry? Is it acceptable to calculate a statistic that *completely* ignores the suspect result(s)? What percentage of the data should be removed by trimming? (In practice 10–25% trimming is common.) Before answering these points, we note the possible use of a simple robust estimate of the standard deviation, based on the interquartile range (IQR, see Section 6.2 above). For a normal error distribution, the IQR is ca. 1.35σ. This relationship clearly supplies a standard deviation estimate which again is not affected by any value taken by the largest or smallest measurements. Unfortunately, the IQR is not a very meaningful concept for very small data sets. Moreover, and somewhat surprisingly, there are several different conventions for its calculation: for large samples, the convention chosen makes little difference, but for small samples the differences in IQR values are large, so trimmed estimates of the standard deviation are not practicable.

A more logical approach to robust estimation can be based on the concept of a **distance function**. Suppose we have a series of n results $x_1, ..., x_n$, and we wish to estimate μ, the mean of the 'reliable' results. In the normal way, our estimate of μ, given here the symbol $\hat{\mu}$, is obtained by minimizing the sum of squares (SS) $\Sigma(|x_i - \mu|)^2$. This sum of *squared* terms is the source of the sensitivity of the mean to large errors. The expression $(x - \mu)^2$ is referred to as a distance function, since it measures the distance of a point from μ. Other distance functions such as $|x - \mu|$ are also feasible. In one established method, any measurement for which $|x - \mu|$ exceeds $c\sigma$, where c normally takes the value 1.5 and σ^2 is a robust variance estimate, is effectively downweighted as described below. The robust variance estimate is usefully derived from a statistic related to the unfortunately abbreviated **median absolute deviation** (MAD!), which is calculated from MAD = median $[|x_i - \text{median} (x_i)|]$. The MAD is an extremely useful statistic — one rough method for evaluating outliers (x_0) is to reject them if $\{|x_i - \text{median} (x_i)|]/\text{MAD} > 5$. It can be shown that the MAD/0.6745 is a useful robust estimate of σ (called $\hat{c}$), which can be used unchanged during the iterative estimates of $\hat{\mu}$.

These techniques can be applied to the measurements discussed above (0.380, 0.400, 0.401, 0.403, 0.410, 0.411, 0.413). First it is necessary to calculate the MAD. The median of these numbers is clearly 0.403 (i.e. the fourth of the seven ordered values), so the individual deviations (without regard to their signs) are 0.023, 0.003, 0.002, 0, 0.007, 0.008, and 0.010; in numerical order: 0, 0.002, 0.003, 0.007, 0.008, 0.010, and 0.023. The MAD is the median of these seven numbers, i.e. 0.007, so $\hat{\sigma}$ = MAD/0.6745 = 0.007/0.6745 = 0.0104, and 1.5$\hat{\sigma}$ is 0.0156.

We are now in a position to begin iterative estimates of $\hat{\mu}$. This process is commenced by taking any reasonable estimate for $\hat{\mu}$, and calculating $|x_i - \hat{\mu}|$ values for each measurement. In this example, suppose the initial $\hat{\mu}$-value is the median, 0.403. As we have seen, the individual deviations from this value are (in numerical order, but neglecting their signs) 0, 0.002, 0.003, 0.007, 0.008, 0.010, and 0.023. In the first iteration for $\hat{\mu}$, the original measurements are retained if these deviations from the median are $\leqslant 0.0156$ — this clearly applies to all the deviations listed except the last. In the event that the deviation is >0.0156, the original value in question is *changed* to become $\hat{\mu} - c\hat{\sigma}$ or $\hat{\mu} + c\hat{\sigma}$, depending on whether it was originally below or above the median respectively. In the present example, the value 0.380, which gives rise to the large deviation of 0.023, has to be changed to $\hat{\mu} - c\hat{\sigma}$, i.e. $0.403 - 0.0156 = 0.3874$.

There is thus now a new data set, with the measurement 0.380 in the original set having been replaced by 0.3874. This new set of numbers is called a set of *pseudo-values* ($\tilde{x}_i$), and the calculation is repeated using this new set. The first step is to calculate the mean of the new values (note that although the initial value of $\hat{\mu}$ can be based on the mean or the median or any other sensible estimate, subsequent steps in the iteration always use the mean): this gives the result 0.4036. The individual deviations from this new estimate of $\hat{\mu}$ are in numerical order, without signs: 0.0006, 0.0026, 0.0036, 0.0064, 0.0074, 0.0094, and 0.0162. As expected (since only one measurement was suspicious in the first place) only the last of these deviations exceeds 0.0156, which means that the measurement in question is changed again, from 0.3874 to $(0.4036 - 0.0156) = 0.3880$. The next set of $\tilde{x}_i$ values is thus the same as the previous set, except that the value 0.3874 is replaced by 0.3880. The new mean ($\hat{\mu}$-value) is thus 0.4037. This is so close to the previous value that it is clearly unnecessary to undertake any further iterations: we conclude that a robust estimate of $\hat{\mu}$ is 0.4037, say 0.404. This example is typical in that there is a very rapid convergence of the iterated values of $\hat{\mu}$.

This calculation deserves several comments. The first is that, like so many iterative procedures, it is much more tedious to describe and explain than it is to perform! The second point to note is that in this example we have estimated μ by making some (robust and reasonable) assumptions about $\hat{\sigma}$: there are also methods where the opposite is true, i.e. a robust estimate of precision is obtained if a mean value is known, and yet more methods in which both robust estimates are calculated iteratively and side by side. Lastly it is worth re-emphasizing that these robust methods do not have the worries and ambiguities of outlier tests. In the example just examined, the Q-test (Section 3.3) suggested that the value 0.380 could be rejected as an outlier ($P = 0.95$), whereas the simple MAD-based test given above suggests that it should not: here $[|x_0 - \text{median}(x_i)|]/\text{MAD} = 3.3$, well below the (rough) critical value of 5. Such concerns and contradictions disappear in robust statistics, where the outliers are neither wholly rejected nor accepted unchanged, but accepted in a changed or downweighted form.

The problems caused by possible outliers in regression calculations have been outlined in Sections 5.13 and 6.9, where rejection using a specified criterion and non-parametric approaches respectively have been discussed. It will be apparent that robust approaches will be of value in regression statistics as well as in the statistics of repeated measurements, and there has indeed been a rapid growth of interest in robust regression methods amongst analytical scientists. A summary of two amongst the many approaches developed must suffice.

In the previous section (6.9) it was noted that a single suspect measurement has a considerable effect on the a and b values calculated for a straight line by the normal 'least squares' method, which seeks to minimize the sum of the squares of the y-residuals. This is because, just as in the nitrite determination example given above, the use of squared terms causes such suspect data points to have a big influence on the sum of squares. A clear and obvious alternative is to seek to minimize the *median* of the squared residuals, which will be much less affected by large residuals. This **least median of squares** (LMS) method is very robust: its *breakdown point*, i.e. the proportion of outliers amongst the data that it can tolerate, is 50%, the theoretical maximum value. (If the proportion of 'suspect' results exceeds 50% it clearly becomes impossible to distinguish them from the 'reliable' results.) The LMS method also works well in cases such as that discussed in Section 5.11, where we wish to characterize the straight-line portion of a set of data which are linear near to the origin, but non-linear at higher x and y values. Its disadvantage is that it involves an iterative calculation which converges rather slowly: that is, many iterations are often required before the estimated a and b values become more or less constant.

Another robust approach to regression (and to repeated measurement statistics) is provided by **Winsorization**. This can be regarded as a variant of the trimming method described above. But instead of removing altogether the data yielding the largest positive or negative residuals, such points are reduced in importance by 'moving' them so that their residuals are equal to the next largest (or perhaps the third largest) positive or negative residual respectively. As in trimming, the procedure is applied to the largest positive and negative residuals: the arbitrariness of the procedure is reduced in this case, however, as the change in the residual value for any 'sensible' point is small. Given the availability of the required software, these and other robust regression techniques are sure to find increasing use in analytical chemistry in future.

6.11 THE KOLMOGOROV TEST FOR GOODNESS OF FIT

In Chapter 3 the common statistical problem of 'goodness of fit' was discussed. This problem arises when it is required to test whether a sample of observations might come from a particular distribution, such as a normal distribution. The chi-squared test is very suitable for this purpose when the data are presented as frequencies, though the test is not normally used for fewer than about 50 observations, and is difficult to use with continuous data. In this section, the Kolmogorov method, which is well suited to testing goodness of fit with continuous data, is described. Extensions of the method, which will not be described in detail, allow it to be extended to the comparison of two samples. These modified methods were first discovered by Smirnov, and the series of tests is often known as the Kolmogorov–Smirnov method.

The principle of the Kolmogorov approach is very simple. It involves comparing the cumulative frequency curve of the data to be tested with the cumulative frequency curve of the hypothesized distribution. The concept of the cumulative frequency curve, and its application in conjunction with normal probability paper, was discussed in Chapter 3. When the hypothetical and experimental curves have been drawn, the test statistic is obtained by finding the maximum vertical difference between them, and comparing this value in the usual way with a set of tabulated values. If the experimental data depart substantially from the expected distribution, the two curves will be expected to be widely separated over part of the cumulative frequency diagram: if, however, the data are closely in accord with the expected distribution, the two curves will never be very far apart. In practice, the Kolmogorov method has two common applications — testing for randomness, and testing for normality of distribution — and a simple example of the latter application will illustrate the operation of the procedure.

When the Kolmogorov method is used to test whether a distribution is normal, we must first transform the original data, which might have any values for their mean and standard deviation, into the **standard normal variable**, z. This is done by using the equation:

$$z = (x - \mu)/\sigma \tag{6.11}$$

where the terms have their usual meanings. The transformation is often referred to as converting the data into *the standard form*. Statistical theory shows that z, calculated in this way, is itself normally distributed, and its cumulative distribution function is given in many collections of statistical tables. Equation (6.11) can be used in two ways. In some cases it may be required to test whether a set of data could come from a *particular* normal distribution, of *given* mean and standard deviation. In such a case, the experimental data are transformed directly by using Eq. (6.11), and the Kolmogorov test is performed. More often, it will simply be required to test whether the data might come from *any* normal distribution. In this instance, the mean and the standard deviation are estimated first, by the simple methods of Chapter 2; the data are next transformed by using Eq. (6.11); then the Kolmogorov method is applied Both types of test are illustrated in the following example.

Example. Eight titrations were performed, with the results 25.13, 25.02, 25.11 25.07, 25.03, 24.97, 25.14, and 25.09 ml. Could such results have come from (a) a normal population with mean 25.00 ml and standard deviation 0.05 ml, and (b) from any other normal population?

(a) Here, the first step is to transform the x-values into z-values, by using the relationship $z = (x - 25.00)/0.05$, obtained from Eq. (6.11). The eight results are thus transformed into 2.6, 0.4, 2.2, 1.4, 0.6, -0.6, 2.8, and 1.8. These z-values are arranged in order and plotted as a cumulative distribution function with a step height of 0.125 (i.e. 1/8). (Note that this is not the same calculation as that used in Section 3.12). Comparison with the hypothetical function for z indicates (Fig. 6.2) that the maximum difference is 0.545 at a z-value just below 1.4. To test this value, Table A.12 is used. Table A.12 shows that, for $n = 8$ and $P = 0.05$, the critical value is 0.288, so the null hypothesis can be rejected — the titration values probably do not come from a normal population with mean 25.00 ml and standard deviation 0.05 ml.

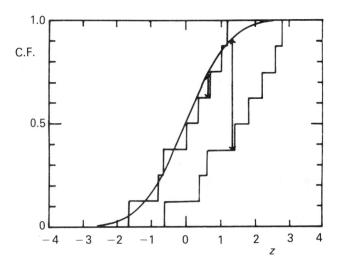

Fig. 6.2. Kolmogorov's method used to test for the normal distribution. Maximum differences between the theoretical cumulative frequency curve and the two tested distributions are shown by the arrows (←→).

(b) In this case, we estimate the mean and the standard deviation [with the aid of Eqs. (2.1) and (2.2)] as 25.07 and 0.059 ml respectively, the latter result being correct to two significant figures. The z-values are now given by $z = (x - 25.07)/0.059$, i.e. by 1.02, -0.85, 0.68, 0, -0.68, -1.69, 1.19, 0.34. The cumulative frequency diagram for these values differs from the hypothetical curve by 0.125 at most (at several points). This difference is much smaller than the critical value of 0.288. We can thus accept the null hypothesis that the data come from a normal population with mean 25.07 and standard deviation 0.059.

6.12 CONCLUSIONS

The rapid and non-parametric tests described in this chapter are only a small fraction of the total available number of such methods. The examples given clearly exemplify their strengths and weaknesses. In many cases their speed and convenience give them a distinct advantage over conventional methods, and the non-parametric tests do not involve assumption of a normal distribution. They are ideally suited to the preliminary examination of small numbers of measurements, and to quick calculations made — often without the need for tables — while the analyst is at the bench or on the shop floor. They can also be used when three or more samples are studied (Section 6.7). The **power** (i.e. the probability that a false null hypothesis is rejected: cf. Section 3.13) of a non-parametric method may be less than that of the corresponding parametric test, but the difference is only rarely serious. For example, many comparisons have been made of the powers in different conditions (i.e. different population distributions and sample sizes) of the Wilcoxon rank sum test and the t-test. The former performs very well in almost all circumstances and is only marginally less powerful than the t-test even when the data come from a normally distributed population. Many suites of programs for personal computers now include several non-parametric tests. Such programs allow a particular set of data to be evaluated rapidly by two or more methods, and seem certain to enhance rather than reduce interest in these very convenient methods.

BIBLIOGRAPHY

W. J. Conover, *Practical Non-parametric Statistics*, Wiley, New York, 1971. Probably the best known general text on non-parametric methods.

W. W. Daniel, *Applied Nonparametric Statistics*, Houghton Mifflin, Boston, 1978. A very comprehensive text, covering a wide range of nonparametric methods in considerable detail; many examples.

R. Langley, *Practical Statistics*, Pan Books, London, 1968. A very readable book, with considerable emphasis on non-parametric tests.

P. Sprent, *Quick Statistics*, Penguin Books, London, 1981. An excellent introduction to non-parametric methods: clear, non-mathematical explanations, and many examples and exercises.

P. J. Rousseeuw and A. M. Leroy, *Robust Regression and Outlier Detection*, John Wiley, New York, 1987. An excellent and remarkably readable book with many illustrative examples.

P. Sprent, *Applied Nonparametric Statistical Methods*, Chapman and Hall, London, 1989. Covers a wide range of significance tests in a practical way, and with a good discussion of robust techniques.

C. Chatfield, *Problem Solving: A Statisticians's Guide*, Chapman and Hall, London, 1988. A thoroughly practical and useful book containing much valuable advice: lengthy appendices summarize the background theory.

P. F. Velleman and D. C. Hoaglin, *Applications, Basics and Computing of Exploratory Data Analysis*, Duxbury Press, Boston, 1991. An excellent introduction to IDA/EDA: computer programs in BASIC and FORTRAN are provided.

EXERCISES

1.   A titration was performed 4 times, with the results: 9.84, 9.91, 9.89, 10.20 ml. Calculate and comment on the median and the mean of these results.

2.   The level of sulphur in batches of an aircraft fuel is claimed by the manufacturer to be symmetrically distributed with a median value of 0.10%. Successive batches are found to have sulphur concentrations of 0.09, 0.12, 0.10, 0.11, 0.08, 0.17, 0.12, 0.14 and 0.11%. Use the sign test, the signed rank test, and a range test to check the manufacturer's claim.

3.   The concentrations (g/100 ml) of immunoglobulin G in the blood sera of ten donors are measured by radial immunodiffusion (r.i.d.) and by electroimmunodiffusion (e.i.d.), with the following results.

Donor:	1	2	3	4	5	6	7	8	9	10
R.i.d. result:	1.3	1.5	0.7	0.9	1.0	1.1	0.8	1.8	0.4	1.3
E.i.d. result.	1.1	1.6	0.5	0.8	0.8	1.0	0.7	1.4	0.4	0.9

Are the results of the two methods significantly different?

4.   Ten carbon rods used successively in an electrothermal atomic-absorption spectrometer were found to last for 24, 26, 30, 21, 19, 17, 23, 22, 25 and 25 samples. Test the randomness of these rod lifetimes.

5.   After each drinking three pints of beer, five volunteers were found to have blood alcohol levels of 104, 79, 88, 120 and 90 mg/100 ml. A further set of six volunteers drank three pints of lager each, and were found to have blood alcohol levels of 68, 86, 71, 79, 91 and 66 mg/100 ml. Use the rank sum test, Tukey's quick test or the Mann–Whitney U-test to investigate the suggestion that drinking lager produces a lower blood alcohol level than drinking the same amount of beer. Use the Siegel–Tukey method to check whether the spreads of these two sets of results differ significantly.

6.   A university chemical laboratory contains seven atomic-absorption spectrometers (A–G). Surveys of the opinions of the research students and of the academic staff show that the student's order of preference for the instruments is B, G, A, D, C, F, E, and that the staff's order of preference is G, D, B, E, A, C, F. Are the opinions of the students and the staff correlated?

7.   Use Theil's method to calculate the regression line for the data of exercise 1 in Chapter 5.

8.   Taking the radial immunodiffusion data from exercise 3 above, use the Kolmogorov method to test the hypothesis that serum immunoglobulin G levels are normally distributed with a mean of 1.0 g/100 ml and a standard deviation of 0.2 g/100 ml. Will any other normal distribution fit the data better?

9.   The nickel levels in three samples of crude oil were determined by atomic-absorption spectrometry, with the following results:

Sample	Measurements (Ni, ppm)					
1	14.2	16.8	15.9	19.1	15.5	16.0
2	14.5	20.0	17.7	18.0	15.4	16.1
3	18.3	20.1	16.9	17.7	17.9	19.3

Use the Kruskal–Wallis method to decide whether the nickel levels in the three oils differ significantly.

7

Experimental design, optimization and pattern recognition

7.1 INTRODUCTION

In Chapter 3 the idea of a factor was introduced. A factor is any aspect of the experimental conditions which affects the value obtained from an experiment. Section 3.9 gave the example of the dependence of a fluorescence signal on the conditions under which a solution was stored. In this experiment the factor of interest was the storage conditions; it is called a controlled factor because it can be altered at will by the experimenter. In another example in the following chapter (Section 4.3), in which salt from different parts of a barrel was tested for purity, the factor of interest, i.e. the part of the barrel from which the salt was taken, was chosen at random. For that reason this factor is called an uncontrolled factor. Both these factors are **qualitative** since their possible 'values' cannot be arranged in numerical order. A factor for which the possible values *can* be arranged in numerical order is **quantitative**, e.g. temperature. The different values which a factor takes are known as different levels.

As presented in Chapters 3 and 4, these experiments were intended as an introduction to the calculations involved in the analysis of variance (ANOVA). No mention was made of the experimental conditions which might have introduced other factors that would affect the results. In the fluorescence experiment such possible factors include: the ambient temperature, whether the same fluorimeter was used for each measurement, and whether the measurements were made by the same person and on the same day. Any of these factors *might* have influenced the results to give the observed behaviour, thus invalidating the conclusion drawn about the effect of the storage conditions. Clearly, if the correct conclusions are to be drawn from an experiment, the various factors affecting the result must be identified and, if possible, controlled. The term **experimental design** is usually used to describe the stages of (1) identifying the factors which may affect the result of an experiment (2) designing the experiment so that the effects of uncontrolled factors are minimized, and (3) using statistical analysis to separate the effects of the various factors involved.

7.2 RANDOMIZATION

One of the assumptions of ANOVA is that the uncontrolled variation is random. However, in measurements made over a period of time, variation in an uncontrolled factor such as pressure, temperature, deterioration of apparatus etc. may produce a trend in the results. As a result the errors due to uncontrolled variation are no longer random since successive errors are correlated. This can lead to a systematic error in the results. Fortunately this problem is simply overcome by using the technique of **randomization**. Suppose we wish to compare the effect of a single factor at three different levels. In an experiment of this type the different levels are often called **treatments**. The term treatment has its origin in the development of ANOVA by Fisher for analysing agricultural experiments. In such cases, different treatments might be, for example, the application of different fertilizers. Let us denote the three different treatments by A, B and C and require that four replicate measurements are made for each treatment. Instead of making the four measurements for treatment A, followed by the four for B, then the four for C, we make the twelve measurements in a random order, decided by using a table of random numbers. Each treatment is assigned a number for each replication:

$$
\begin{array}{ccc}
\text{A} & \text{B} & \text{C} \\
01\ 02\ 03\ 04 & 05\ 06\ 07\ 08 & 09\ 10\ 11\ 12
\end{array}
$$

(Note that each number has the same number of digits). We then enter a random number table (see Table A.6) at an arbitrary point and read off pairs of digits, discarding the pairs 00, 13–99 and repeats. Suppose this gives the sequence 02, 10, 04, 03, 11, 01, 12, 06, 08, 07, 09, 05, then the measurements would be made at the different levels in the following order A, C, A, A, C, A, C, B, B, B, C, B. This random order of measurement ensures that the errors at level A due to uncontrolled factors are random, likewise for levels B and C.

7.3 BLOCKING

One disadvantage of complete randomization is that it fails to take advantage of any natural subdivisions in the experimental material. Suppose, for example, all the twelve measurements in the previous example could not be made on the same day but had to be divided between four consecutive days. Using the same order as before would give:

Day 1	A	C	A
Day 2	A	C	A
Day 3	C	B	B
Day 4	B	C	B

With this design the measurements using treatment A occur on the first two days whereas those using treatment B are made on the last two days. If it seemed that there was a difference between treatments A and B it would not be possible to tell whether this difference was genuine or was caused by the effect of using the two treatments on different pairs of days. A better design is one in which each treatment is used once on each day, with the order of the treatments randomized on each day. For example:

$$
\begin{array}{llll}
\text{Day 1} & \text{A} & \text{C} & \text{B} \\
\text{Day 2} & \text{A} & \text{B} & \text{C} \\
\text{Day 3} & \text{C} & \text{B} & \text{A} \\
\text{Day 4} & \text{C} & \text{A} & \text{B}
\end{array}
$$

A group which contains one measurement for each treatment is known as a **block**, (again a term of agricultural origin signifying an area of a field) and the design is called a **randomized block** design. In this case the measurements on each day form a block.

Another advantage of a blocked experiment is that it is more sensitive than an unblocked one as a method of testing whether different treatments produce significantly different results. ANOVA — in this case two-way because each measurement is classified according to two factors, treatment and day — allows the separation of the three sources of variation: between-block variation, between-treatment variation and random variation due to experimental error. In order to decide whether the treatments produce significantly different results, the between-treatment estimate of variance is compared with the variance estimated from the random errors. The sensitivity of the experiment depends on the size of the random variation: the smaller the random variation, the smaller the difference between the treatments which can be detected. In an unblocked experiment the random variation would be larger since it would include a contribution due to day-to-day variation and consequently the sensitivity of the experiment would be reduced.

7.4 TWO-WAY ANOVA

In two-way ANOVA each measurement, x_{ij}, is classified according to two factors as shown in Table 7.1. There are N measurements divided between c treatment levels

Table 7.1.   General form of table for two-way ANOVA

	Treatment					Row total
	1	2	.....	j	c	
Block 1	x_{11}	x_{12}		x_{1j}	x_{1c}	$T_1.$
Block 2	x_{21}	x_{22}		x_{2j}	x_{2c}	$T_2.$
.	.	. .....		......	.	.
Block i	x_{i1}	x_{i2}		x_{ij}	x_{ic}	$T_i.$
.	.	. .....		......	.	.
Block r	x_{r1}	x_{r2}		x_{rj}	x_{rc}	$T_r.$
Column total	$T._1$	$T._2$		$T._j$	$T._c$	T = grand total

and r blocks (therefore $N = cr$). The column and row totals and the grand total, T, are also given, as these are needed in the calculation. The derivation of the formulae used will not be given in detail as they were for one-way ANOVA in Section 3.9. The principle of the method is the same and references are given at the end of the chapter for those who wish to pursue the mathematics. The formulae for calculating the variation from the three different sources, viz. between-treatment, between-block and experimental error, are given in Table 7.2.

Table 7.2.   Formulae for two-way ANOVA

Source of variation	Sum or squares	Degrees of freedom
Between-treatment	$\sum_{j=1}^{c} T_j^2/r - T^2/N$	$c - 1$
Between-block	$\sum_{i=1}^{r} T_i^2/c - T^2/N$	$r - 1$
Residual	by subtraction	by subtraction
Total	$\sum_i \sum_j x_{ij}^2 - T^2/N$	$N - 1$

Example. In an experiment to compare the percentage efficiency of different chelating agents in extracting metal ions from aqueous solution the following results were obtained:

	Chelating agent			
Day	A	B	C	D
1	84	80	83	79
2	79	77	80	79
3	83	78	80	78

On each day a fresh solution of the metal ion (with the specified concentration) was prepared and the extraction performed with each of the chelating agents taken in a random order.

In this experiment the use of different chelating agents is a controlled factor since the chelating agents are chosen by the experimenter; the day on which the experiment is performed introduces uncontrolled variation caused both by changes in laboratory temperature, pressure etc. and slight differences in the concentration of the metal ion solution, i.e. the day is a random factor. In Chapters 3 and 4 it was explained how ANOVA can be used either to test for a significant effect due to a controlled factor, or to estimate the variance of an uncontrolled factor. In this experiment, where there are both types of factor, two-way ANOVA can be used in both ways: (i) to test whether the different chelating agents have significantly different efficiencies, and (ii) to test whether the day-to-day variation is significantly greater than the variation due to the random error of measurement and, if it is, to estimate its variance.

As in one-way ANOVA, the calculations can be simplified by subtracting an arbitrary number from each measurement. The table below shows the measurements with 80 subtracted from each.

Blocks	Treatments				Row totals, $T_{i.}$	$T_{i.}^2$
	A	B	C	D		
Day 1	4	0	3	−1	6	36
Day 2	−1	−3	0	−1	−5	25
Day 3	3	−2	0	−2	−1	1
Column totals, $T_{.j}$	6	−5	3	−4	0 = Grand total, T	
$T_{.j}^2$	36	25	9	16		

$62 = \sum_i T_{i.}^2$

$$\sum_j T_{.j}^2 = 86$$

We have: $r = 3$, $c = 4$, $N = 12$, $\sum_i T_{i.}^2 = 62$, $\sum_j T_{.j}^2 = 86$, $T = 0$, $\sum_i \sum_j x_{ij}^2 = 54$.

The calculation of the mean square for each source of variation is shown in the next table.

Source of variation	Sum of squares	Degrees of freedom	Mean square
Between-treatment	$86/3 - 0^2/12 = 28.6667$	3	$28.6667/3 = 9.5556$
Between-block	$62/4 - 0^2/12 = 15.5$	2	$15.5/2 = 7.75$
Residual	by subtraction = 9.8333	6	$9.8333/6 = 1.6389$
Total	$54 - 0^2/12 = 54.0$	11	

(Since the residual sum of squares involves calculating a difference, as many significant places as possible should be carried in order to avoid errors in this quantity.)

The reader may find it instructive to verify that the sums of squares *do* separate the treatment and block effects. This can be done by (say) increasing all the values in one block by a fixed number and recalculating the sums of squares. It should be found that although the total and between-block sums of squares are altered, the between-treatment and residual sums of squares are unchanged.

If there is no difference between the efficiencies or days then all three mean sums of squares should give an estimate of σ_0^2, the variance of the random variation due to experimental error (see Section 3.9). The F-test is used to see whether the estimates of variance differ significantly. Comparing the between-treatment mean square with the residual mean square gives:

$$F_{3,6} = 9.5556/1.6389 = 5.83$$

From Table A.2 the critical value of F is 4.76 ($P = 0.05$) indicating that there is a significant difference between treatments, i.e. between the efficiency of different chelating agents, at the 5% level. Comparing the between-block (i.e. between-day) and residual mean squares gives:

$$F_{2,6} = 7.75/1.6389 = 4.76$$

The critical value of F is 5.143 ($P = 0.05$) so there is no significant difference between days. Nevertheless the between-block mean square is considerably larger than the residual mean square, and had the experiment been unblocked, so that these two effects were combined in the estimate of experimental error, the experiment would probably have been unable to detect whether different treatments gave significantly different results. If the difference between days *had* been significant it would indicate that other factors such as temperature, pressure, the preparation of the solution etc. were having an effect. It can be shown that the between-block mean square gives an estimate of $\sigma_0^2 + c\sigma_b^2$ where σ_b^2 is the variance of the random day-to-day variation. Since the residual mean square gives an estimate of σ_0^2, an estimate of σ_b^2 can be deduced.

This example illustrates clearly the benefits of considering the design of an experiment before it is performed. Given a blocked and an unblocked experiment with the same number of measurements in each, the former is more sensitive and yields more information.

The analysis above is based on the assumption that the effects, if any, are additive. This point is discussed further in Section 7.7.

7.5 LATIN SQUARES

An uncontrolled factor which was not taken into account in the analysis of the previous section was the time of day at which the measurements were made. Systematic variation during the day due to deterioration of the solution or one-directional change in laboratory temperature could produce a trend in the results. It is possible to use an experimental design which allows the separation of this factor if there are equal numbers of blocks and treatments. Instead of, for example, the following randomized block design

Day 1	A	C	B
Day 2	C	A	B
Day 3	C	B	A

where measurements using treatment C are concentrated at the beginning of the day and those using B are concentrated at the end of the day, the following design could be used:

Day 1	A	B	C
Day 2	C	A	B
Day 3	B	C	A

This design, in which each treatment appears once in each row and once in each column, is known as a **Latin square**; it allows the separation of the variation into between-treatment, between-day, between-time-of-day, and random experimental error.

If the square is greater than 3×3 there is more than one possible design and one of these designs should be chosen at random. The obvious disadvantage of a simple Latin square design is that it requires equal numbers of blocks and treatments. More complicated designs are possible which remove this constraint.

7.6 NESTED AND CROSS-CLASSIFIED DESIGNS

The design of experiment given in Section 7.4 is known as **cross-classified** because a measurement is made for each possible combination of the different levels of the factors. Contrast it with the experimental design shown in Fig. 7.1, where a sample of

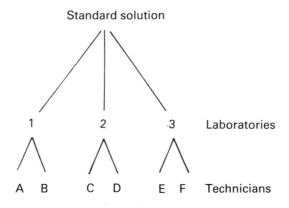

Fig. 7.1.   Nested or hierarchical design.

a standard solution is sent to each of three laboratories and in each laboratory the sample is analysed by two different technicians. Such a design is is said to be **nested** or **hierarchical** and it can be seen that not all possible combinations of experimental conditions are used, e.g. the technicians do not make measurements in laboratories other than their own. (One-way ANOVA can also be considered as a nested design.) Mixed (i.e. partly cross-classified, partly nested) designs are also possible.

7.7 INTERACTION

A basic assumption of the analysis performed in Section 7.4 was that the effects of the two factors were **additive**. The meaning of this term is most easily explained by a numerical example. The simplest situation is that in which each factor has two levels: e.g. factor A has levels a_1 and a_2 and factor B has levels b_1 and b_2. Suppose, for example, the results for the combinations a_1b_1, a_1b_2 and a_2b_1 are 10, 15 and 12 as shown in Table 7.3. If these are the true values, i.e. there is no random variation,

Table 7.3

| | | Level of A | |
		a_1	a_2
Level of B	b_1	10	12
	b_2	15	?

then the effect of changing factor B from level b_1 to b_2 with A at a_1 is 5. If the effects are additive then the effect of changing B from level b_1 to level b_2 with A at a_2 should also be 5, so the final value in the table should be 17. Note that the effect of changing A from level a_1 to level a_2 is then 2, irrespective of the level of B. In general, if the results are as shown in Table 7.4, and they are free from random error, then $y_2 - y_1$

Table 7.4

| | | Level of A | |
		a_1	a_2
Level of B	b_1	y_1	y_2
	b_2	y_3	y_4

equals $y_4 - y_3$ if the effects are additive. If the effects are not additive, we say there is **interaction** between A and B because a particular combination of A and B gives a higher value than expected. Figure 7.2 illustrates the effect of interaction graphi-

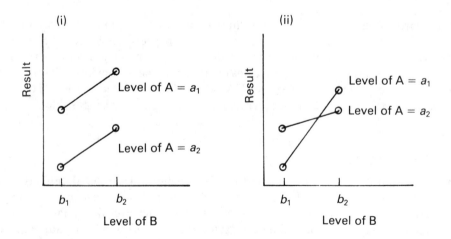

Fig. 7.2.   (i) Effects which are additive; (ii) effects which interact.

cally. On both graphs each result is plotted against the level of B, and points with the same level of A are joined. If the effects are additive then the lines will be parallel as in the left-hand diagram; if the effects interact, the lines will not be parallel, as in the right-hand diagram. If both factors have several levels, a graph of this type is helpful in the interpretation of any interaction.

Unfortunately the experimental situation is confused by the presence of random errors. The reader may check that for Table 7.3, two-way ANOVA gives zero for the residual sum of squares if the missing value is taken as 17, but if one of the values is altered this is no longer so. With this design of experiment we cannot assess the extent to which a non-zero residual sum of squares is caused by random error, rather than by interaction between the two effects. In order to estimate the random error the measurements in each cell must be repeated. The exact form of this repetition is important. Suppose, for example, a determination involves dissolving a given weight of a specimen in a specified volume of water and titrating against a standard solution. The repeated measurements should involve both the weighing and titration stages: if only the titration stage is repeated the random error in weighing would not be included in the estimate of experimental error. Further, if the same batch of standard solution and the same set of glassware are not used for *all* the determinations in the experiment then the same batch of standard solution and the same set of glassware must *not* be used for repeat measurements. Measurements which are repeated so that they are subject to all the sources of random error in the experiment are called **replicates**. The method by which the random error and interaction sums of squares can then be separated is illustrated by the following example.

Example. In an experiment to investigate the validity of a solution as a liquid absorbance standard, the value of the molar absorptivity, ε, of solutions of three different concentrations was measured at four different wavelengths. Two replicate measurements were made for each combination of concentration and wavelength. The order in which the measurements were made was randomized. The results are shown in Table 7.5: for simplicity of calculation the measured values have been multiplied by 100.

Table 7.5

Concentration (g/l.)＼λ (nm)	240	270	300	350
0.02	94, 96	106, 108	48, 51	78, 81
0.06	93, 93	106, 105	47, 48	78, 78
0.10	93, 94	106, 107	49, 50	78, 79

The first stage of the calculation is to find the cell totals. This is done in Table 7.6, which also includes other quantities needed in the calculation. As before, T_i.

Table 7.6

Concentration (g/l.) \ λ(nm)	240	270	300	350	$T_{i\cdot}$	$T_{i\cdot}^2$
0.02	190	214	99	159	662	438244
0.06	186	211	95	156	648	419904
0.10	187	213	99	157	656	430336

$T_{\cdot j}$ 563   638   293   472 $T = 1966$ 1288484 $= \sum_i T_{i\cdot}^2$

$T_{\cdot j}^2$ 316969   407044   85849   222784

$$\sum_j T_{\cdot j}^2 = 1032646$$

denotes the total of the ith row, $T_{\cdot j}$ the total of the jth column and T the grand total.

As before, the between-row, between-column and total sums of squares are calculated. Each calculation requires the term T^2/nrc (where n is the number of replicate measurements in each cell, in this case 2, r = number of rows and c = number of columns). This term is sometimes called the **correction term**, C. We have:

$$C = T^2/nrc = 1966^2/(2 \times 3 \times 4) = 161048.16$$

The sums of squares are now calculated:

$$\text{Between-row sum of squares} = \sum_i T_{i\cdot}^2/nc - C$$

$$= 1288484/(2 \times 4) - 161048.16$$
$$= 12.34$$

with $r - 1 = 2$ degrees of freedom

$$\text{Between-column sum of squares} = \sum_j T_{\cdot j}^2/nr - C$$

$$= 1032646/(2 \times 3) - 161048.16$$
$$= 11059.506$$

with $c - 1 = 3$ degrees of freedom.

$$\text{Total sum of squares} = \sum x_{ijk}^2 - C$$

where x_{ijk} is the kth replicate in the ith row and jth column, i.e. $\sum x_{ijk}^2$ is the sum of the squares of the individual measurements in Table 7.5.

$$\text{Total sum of squares} = 172138 - 161048.16$$
$$= 11089.84$$

with $nrc - 1 = 23$ degrees of freedom.

The variation due to random error (usually called the **residual variation**) is estimated from the within-cell variation, i.e., the variation between replicates.

The residual sum of squares $= \Sigma x^2_{ijk} - \Sigma T^2_{ij}/n$ where T_{ij} is the total for the cell in the ith row and jth column, i.e. the sum of the replicate measurements in the ith row and jth column.

$$\text{Residual sum of squares } = \Sigma x^2_{ijk} - \Sigma T^2_{ij}/n$$
$$= 172138 - 344244/2$$
$$= 16$$

with $(n-1)rc=12$ degrees of freedom.

The interaction sum of squares and number of degrees of freedom can now be found by subtraction. The results of these calculations are summarized below.

Source	Sum of squares	Degrees of freedom	Mean square
Between-row	12.34	2	6.17
Between-column	11059.506	3	3686.502
Interaction	1.994	6	0.3323
Residual	16	12	1.3333
Total	11089.84	23	

Each source of variation is compared with the residual mean square to test whether it is significant.

(i) **Interaction**. This is obviously not significant since the interaction mean-square is less than the residual mean square.

(ii) **Between-column** (i.e. between-wavelength). This is significant since we have:

$$F_{3,12} = 3686.502/1.3333 = 2765$$

The critical value of $F_{3,12}$ is 3.49 ($P = 0.05$). In this case a significant result would be expected since absorbance is wavelength-dependent.

(iii) **Between-row** (i.e. between-concentration). We have:

$$F_{2,12} = 6.17/1.3333 = 4.63$$

The critical value of $F_{2,12}$ is 3.885 ($P = 0.05$), indicating that the between-row variation is too great to be accounted for by random variation: the solution is not suitable as an absorbance standard. Figure 7.3 shows the molar absorptivity plotted against wavelength, with the values for the same concentration joined by straight lines. This illustrates the results of the analysis above in the following ways:

(i) the lines are parallel, indicating no interaction;

(ii) the lines are not quite horizontal, indicating that the molar absorptivity varies with concentration;

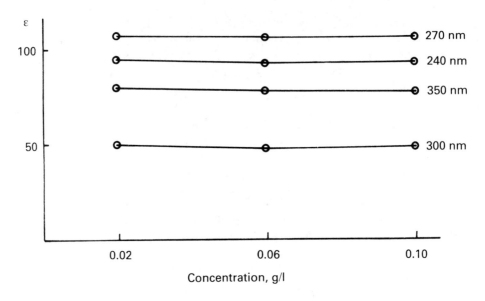

Fig. 7.3.   Relationships in the two-way ANOVA example of Section 7.7.

(iii) the lines are at different heights on the graph, indicating that the molar absorptivity is wavelength-dependent.

The formulae used in the calculation above are summarized in Table 7.7.

Table 7.7.   Formulae for two-way ANOVA with interaction

Source of variation	Sum of squares	Degrees of freedom
Between-row	$\sum_i T_{i.}^2/nc - C$	$r - 1$
Between-column	$\sum_j T_{.j}^2/nr - C$	$c - 1$
Interaction	by subtraction	by subtraction
Residual	$\sum x_{ijk}^2 - \sum T_{ij}^2/n$	$rc(n - 1)$
Total	$\sum x_{ijk}^2 - C$	$rcn - 1$

In this experiment both factors, i.e. the wavelength and the concentration of the solution, are controlled factors. In analytical chemisty an important application of ANOVA is the investigation of two or more controlled factors and their interactions in optimization experiments. This is discussed in Section 7.9.

As discussed in Section 4.8, another important application of ANOVA is in collaborative investigations of precision and accuracy between laboratories. A more elaborate version of the experiments described in Section 4.8 would be one in which several different types of sample were sent to a number of laboratories with each laboratory performing a number of replicate analyses on each sample. Mathematical analysis of the results would yield the following sums of squares: between-laboratory, between-samples, laboratory–sample interaction, and residual. The purpose of such an experiment would be to test first whether there is any interaction between laboratory and sample, i.e. whether some laboratories showed unexpectedly high or low results for some samples. This is done by comparing the interaction and residual sums of squares. If there is *no* interaction, then we could test whether the laboratories obtained significantly different results, i.e. if there is any systematic difference between laboratories. If there is, then the interlaboratory variance can be estimated. However, if there *is* a significant interaction, the testing for a significant difference between laboratories has little relevance.

For two-way ANOVA to be valid the following conditions must be fulfilled (see Section 3.9):

(i) the random error is the same for all combinations of the levels of the factors;
(ii) the random errors are approximately normally distributed.

7.8 FACTORIAL VERSUS ONE-AT-A-TIME DESIGN

An experiment such as the one in the previous example, where the response variable (i.e. the molar absorptivity) is measured for all possible combinations of the chosen factor levels is known as a **complete factorial design**. The reader may have noticed that this design of the experiment is the antithesis of the classical approach in which the response is investigated for each factor in turn while all the other factors are held at a constant level. There are two cogent reasons for using a factorial design rather than a classical design in experiments which test whether the response depends on factor level:

(i) the factorial experiment detects and estimates any interaction, which the one-at-a-time experiment cannot do;

(ii) if the effects of the factors are additive, then the factorial design needs fewer measurements than the classical approach in order to give the same precision. This can be seen by turning again to the molar absoptivity experiment. There, all 24 measurements were used to estimate the effect of varying the wavelength and the *same* 24 were used to estimate the effect of varying concentration. In a one-at-a-time experiment, first the concentration would have been fixed and, to obtain the same precision for the effect of varying the wavelength, 6 measurements would have been needed at each wavelength, i.e. 24 in all. Then the wavelength would have been fixed and another 24 measurements made at different concentrations, making a total of 48 altogether. In general, for k factors, a classical approach involves k times as many measurements as a factorial one with the same precision.

7.9 FACTORIAL DESIGN AND OPTIMIZATION

Since analytical techniques are frequently concerned with detecting minute traces of analyte, it is usually important that the levels of the factors on which the response depends are chosen so as to maximize this response. The process of finding these optimum factor levels is known as **optimization**. The first step is to determine which factors and which interactions between them are important in affecting the response. This can be done by using a factorial experiment with each factor at two levels usually known as 'low' and 'high'. In the case of a quantitative variable the terms 'low' and 'high' have their usual meaning. The exact choice of levels is determined principally by the experience and knowledge of the experimenter and the physical constraints of the system, e.g. if water is used as the solvent only temperatures in the range 0–100°C are practicable. Some problems attendant on the choice of levels are discussed below. For a qualitative variable 'high' and 'low' refer to a pair of different conditions, e.g. the presence or absence of a catalyst, use of mechanical or magnetic stirring, a sample in powdered or granular form, etc. Since we have already considered two-factor experiments in some detail we will turn to one with three factors: A, B and C. This means that there are $2 \times 2 \times 2 = 8$ possible combinations of factor levels, as shown in the table below. A plus sign denotes that the factor is at the high level and a minus sign that it is at the low level. The first column gives a notation often used to describe the combinations, where the presence of the appropriate lower case letter indicates that the factor is at the high level and its absence that the factor is at the low level. The number 1 is used to indicate that all factors are at the low level.

Combination	A	B	C	Response
1	–	–	–	y_1
a	+	–	–	y_2
b	–	+	–	y_3
c	–	–	+	y_4
bc	–	+	+	y_5
ac	+	–	+	y_6
ab	+	+	–	y_7
abc	+	+	+	y_8

The method by which the effects of the factors and their interactions are estimated is illustrated by the following example.

Example. In a high-performance liquid chromatography experiment, the dependence of the retention parameter, k', on three factors was investigated. The factors were pH (factor P), the concentration of a counter-ion (factor T) and the concentration of the organic solvent in the mobile phase (factor C). Two levels were used for each factor and two replicate measurements made for each combination. The measurements were randomized. The table below gives the average value for each pair of replicates.

Combination of factor levels	k'
1	4.7
p	9.9
t	7.0
c	2.7
pt	15.0
pc	5.3
tc	3.2
ptc	6.0

(a) **Effect of individual factors.**

The effect of changing the level of P can be found from the average difference in response when P changes from high to low level with the levels of C and T fixed. There are four pairs of responses which give an estimate of the effect of the level of P as shown in the table below.

Level of C	Level of T	Level of P +	Level of P –	Difference
–	–	9.9	4.7	5.2
+	–	5.3	2.7	2.6
–	+	15.0	7.0	8.0
+	+	6.0	3.2	2.8
				Total = 18.6

Average effect of altering the level of P = 18.6/4 = 4.65

The average effect of altering the levels of T and C can be found in a similar way. It is left as an exercise for the reader to show that they are:

average effect of altering the level of C = −4.85

average effect of altering the level of T = 2.15

(b) **Interaction between two factors.**

Consider now the two factors P and T. If there is *no* interaction between them then the change in response between the two levels of P should be independent of the level of T. The first two figures in the last column of the table above give the change in response when P changes from high to low level with T at low level. Their average is (5.2 + 2.6)/2 = 3.9. The last two figures in the same column give the effect of changing P when T is at high level. Their average is (8.0 + 2.8)/2 = 5.4. If there is no interaction and no random error (see Section 7.7) these estimates of the effect of changing the level of P should be equal. The convention is to take half their difference as a measure of the interaction:

effect of PT interaction = (5.4−3.9)/2 = 0.75.

It is important to realize that this quantity estimates the degree to which the effects of P and T are not additive. It could equally well have been calculated by considering how far the change in response for the two levels of T is independent of the level of P.

The other interactions are calculated in a similar fashion. Again the reader may check that:

effect of CP interaction = −1.95

effect of CT interaction = −1.55

(c) **Interaction between three factors.**

The PT interaction calculated above can be split into two parts according to the level of C. With C at low level the estimate of interaction would be $(8.0 - 5.2)/2 = 1.4$ and with C at high level it would be $(2.8 - 2.6)/2 = 0.1$. If there is no interaction between all three factors and no random error, these estimates of the PT interaction should be equal. The three-factor interaction is estimated by half their difference $[= (0.1 - 1.4)/2 = -0.65]$. The three-factor interaction measures the extent to which the effect of the PT interaction and the effect of C are not additive: it could equally well be calculated by considering the difference between the PC estimates of interaction for low and high levels of T or the difference between the TC estimates of interaction for low and high levels of P.

These results are summarized in the table below.

	Effect
Single factor (main effect)	
P	4.65
T	2.15
C	−4.85
Two-factor interactions	
TP	0.75
CT	−1.55
CP	−1.95
Three-factor interactions	
PTC	−0.65

These calculations have been presented in some detail in order to make the principles clear. An algorithm due to Yates (see references) simplifies the calculation.

In order to test which effects, if any, are significant, ANOVA may be used (provided that there is homogeneity of variance). It can be shown that in a *two*-level experiment, like this one, the required sums of squares can be calculated from the estimated effects by using

$$\text{Sum of squares} = N \times (\text{estimated effect})^2/4$$

where N is the total number of measurements, including replicates. In this case N is 16 since two replicate measurements were made for each combination of factor levels. The calculated sums of squares are given below.

Factor(s)	Sum of squares
P	86.49
T	18.49
C	94.09
PT	2.25
TC	9.61
PC	15.21
PTC	1.69

It can be shown that each sum of squares has one degree of freedom and since the mean square is given by

$$\text{mean square} = \text{sum of squares/number of degrees of freedom}$$

each mean square is simply the corresponding sum of squares. To test for the significance of an effect, the mean square is compared with the error (residual) mean square. This is calculated from the individual measurements by the method described in the molar absorptivity example in Section 7.7. In the present experiment the calculated residual mean square was 0.012 with 8 degrees of freedom. Testing for significance, starting with the highest order interaction, we have for the PTC interaction:

$$F_{1,8} = 1.69/0.012 = 141$$

which is obviously significant. If there is interaction between all three factors there is no point in testing whether the factors taken in pairs or singly are significant, since all factors will have to be considered in any optimization process. A single factor should only be tested for significance if it does not interact with other factors.

One problem with a complete factorial experiment such as this is that the number of experiments required rises rapidly with the number of factors: for k factors at 2 levels with 2 replicates for each combination of levels, 2^{k+1} experiments are necessary e.g. for 5 factors, 64 experiments. When there are more than three factors some economy is possible by assuming that three-way and higher order interactions are negligible. The sums of squares corresponding to these interactions can then be combined to give an estimate of the residual sum of squares, and replicate measurements are no longer necessary. The rationale for this approach is that higher order effects are usually much smaller than main effects and two-factor interaction effects. If higher order interactions *can* be assumed negligible, a suitable fraction of all possible combinations of factor levels is sufficient to provide an estimate of the main and two-factor interaction effects. As mentioned in Section 4.7, such an experimental design is called an **incomplete** or **fractional factorial design**.

Another problem in using a factorial design to determine which factors have a significant effect on the response is that, for factors which are continuous variables, the effect depends on the high and low levels used. If the high and low levels are too close together the effect of the corresponding factor may be found not significant despite the fact that over the whole possible range of factor levels the effect of this factor is *not* negligible. On the other hand, if the levels are further apart they may fall on either side of a maximum and still give a difference in response which is not significant.

7.10 ALTERNATING VARIABLE SEARCH METHOD OF OPTIMIZATION

When the factors which are important in affecting the response have been determined the next step is to alter their levels to obtain the maximum response. The most obvious method is to adopt a one-at-a-time approach, varying the factors separately. The method can be demonstrated graphically when the response depends on two factors which are continuous variables. In this case the relationship between the response and the levels of the two factors can be represented by a surface in three dimensions as shown in Fig. 7.4. This surface is known as the **response surface**. It is more conveniently shown as a contour diagram (Fig. 7.5), where the response on each contour is constant. The form of the contour lines is, of course, unknown to the experimenter who wishes to determine the optimum levels, x_0 and y_0, for the

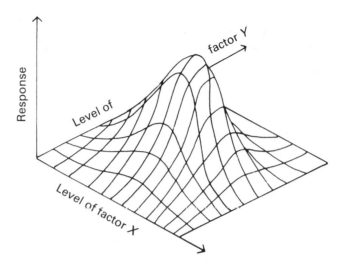

Fig. 7.4.   A response surface for two factors.

factors X and Y respectively. Adopting a one-at-a-time approach with the initial level of X fixed at x_1, say, and varying the level of Y would give a maximum response at the point A, where the level of Y is y_1. Next, holding the level of Y at y_1 and varying the level of X would give a maximum at B. Obviously this is not the true maximum, as the position obtained depends on the initial value chosen for x_1. A better response can be obtained by repeating the process, varying the levels of X and Y alternately. This method is known as the **alternating variable search** or sometimes the **iterative univariate method**. It is relatively straightforward if the response can be monitored continuously as the factor level is altered, for example in spectrometry when the monochromator wavelength is scanned. If such monitoring is not possible then a choice of step size must be made for the change in each of the factors: a more sophisticated method would allow alteration in this step size depending on the observed change in response.

When there is no interaction between the two factors this method is extremely efficient. In such a case the response surface has the form of Fig.7.7(i) or (ii) and varying X and then Y just once will lead to the maximum response. If, however, there is interaction between the two variables then the response surface has the form of Fig. 7.5 and X and Y must then be varied in turn a number of times. In some cases, even this will not lead to the true maximum: this is illustrated in Fig. 7.6 where, although C is not the true maximum, the response falls on either side of it in both the X and the Y directions. A one-at-a-time method arriving at this point would therefore conclude that it represented the maximum response.

There are obviously some problems in applying a one-at-a-time approach to optimization, and a number of other methods have been developed in an attempt to overcome these problems and to minimize the number of experiments required. All these methods can be applied to any number of factors, but the response surface cannot easily be visualized for three or more factors: our remaining discussion of optimization methods will thus largely be confined to experiments involving two factors.

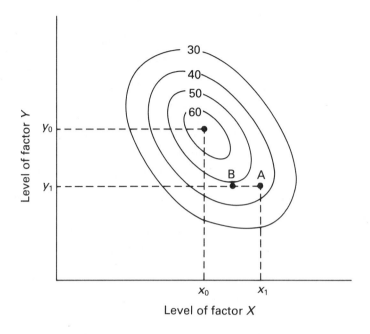

Fig. 7.5.   The contour diagram for a two-factor response surface.

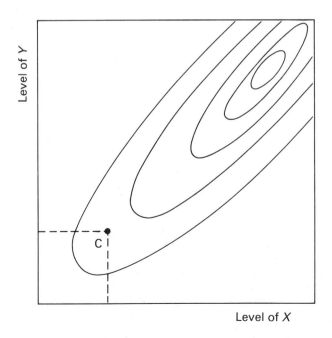

Fig. 7.6.   Contour diagram: a situation in which the one-at-a-time method fails to locate the maximum.

7.11   METHOD OF STEEPEST ASCENT

The process of optimization can be visualised in terms of a person standing on a hill in thick fog with the task of finding the summit! In these circumstances an obvious approach is to walk in the direction in which the hill is steepest. This is the basis of the **method of steepest ascent**. Figure 7.7 shows two possible contour maps. A little consideration shows that the direction of steepest ascent at any point is at right angles to the contour lines at that point, as indicated by the arrows. When the contour lines are circular this will be towards the summit but when the contour lines are elliptical it may not. The shape of the contour lines depends on the scales chosen for the axes: the best results are obtained from the method if the axes are scaled so that a unit change in either direction gives a roughly equal change in response. The first step is to perform a factorial experiment with each factor at two levels. The levels are chosen so that the design forms a square as shown in Fig. 7.8. Suppose, for example, that the experiment is an enzyme catalysed reaction in which the reaction rate, which in this case is the response, is to be maximized with respect to pH (X) and temperature (Y). The table below gives the results (reaction rate measured in arbitrary units) of the initial factorial experiment.

		pH (X)	
		6.8	7.0
Temperature (°C)	20	30	35
	25	34	39

The effects of the two factors can be separated as described in Section 7.9. Rewriting the table above, in the notation of that section, gives:

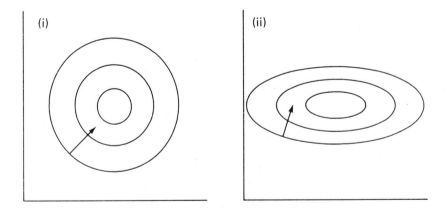

Fig. 7.7.   Contour diagrams: the arrow in each diagram indicates the path of steepest ascent.
In (i) it goes close to the maximum but in (ii) it does not.

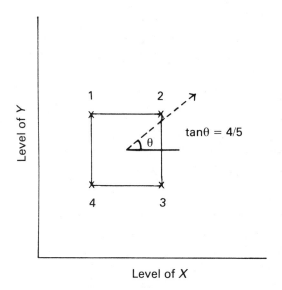

Fig. 7.8.   A 2×2 factorial design to determine the direction of steepest ascent, indicated by
the broken line.

Combination of levels	Rate of reaction
1	30
x	35
y	34
xy	39

Average effect of change in level of X = $[(35-30)+(39-34)]/2 = 5$
Average effect of change in level of Y = $[(34-30)+(39-35)]/2 = 4$

The effects of X and Y indicate that in Fig. 7.8 we should seek for the maximum response to the right and above the original region. Since the change in the X direction is greater than in the Y direction the distance moved in the former should be greater. To be more exact, the distance moved in the X direction should be in the ratio 5:4 to the direction moved in the Y direction, i.e. in the direction indicated by the dotted line in Fig. 7.9.

The next step in the optimization is to carry out further experiments in the direction indicated by the dotted line in Fig. 7.9, at (say) the points numbered 5, 6 and 7. This would indicate point 6 as a rough position for the maximum in this direction. Another factorial experiment is carried out in the region of point 6 to determine the new direction of steepest ascent.

This method gives satisfactory progress towards the maximum provided that, over the region of the factorial design, the contours are approximately straight. This is equivalent to the response surface being a plane which can be described mathematically by a linear combination of terms in x and y. Nearer the summit terms in xy, x^2 and y^2 are also needed to describe the surface. The xy term represents the interaction

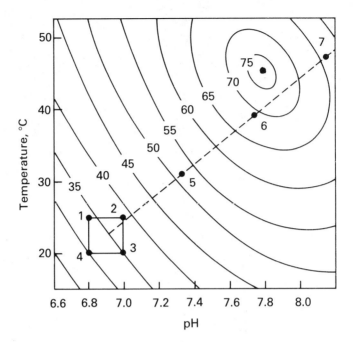

Fig. 7.9.   Contour diagram: the initial direction of steepest ascent is shown by the broken line. Further experiments are done at points 5, 6 and 7.

between X and Y and can be estimated by using replication as described in Section 7.9. The squared terms, which represent the curvature of the surface, can be estimated by comparing the response at the centre of the factorial design with the average of the responses at the corners. When interaction and curvature effects become appreciable compared with the experimental error (which is estimated by replication) a more elaborate factorial design is used which allows the form of the curved surface to be determined and hence the approximate position of the maximum.

It is evident that factorial design and the method of steepest ascent will be very complicated when several factors are involved. The next section describes a method of optimization which is conceptually much simpler.

7.12 SIMPLEX OPTIMIZATION

Simplex optimization may be applied when all the factors are continuous variables. (The reader should be warned that the choice of name is unfortunate since the method is different from the simplex method used in linear programming.) A **simplex** is a geometrical figure which has $n + 1$ vertices when a response is being optimized with respect to n factors. For example, for two factors the simplex will be a triangle. The method of optimization is illustrated by Fig.7.10. The initial simplex is defined by the points labelled 1, 2 and 3. In the first experiment the response is measured at

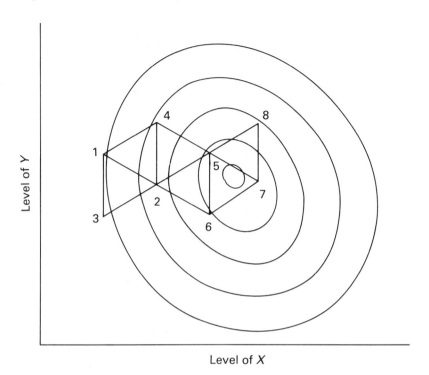

Fig. 7.10.   Simplex optimization.

each of the combinations of factor levels given by the vertices of the simplex. The worst response in this case would be found at point 3 and it would be logical to conclude that a better response would be found at a point which is the reflection of 3 with respect to the line joining 1 and 2, i.e. at 4. The points 1, 2 and 4 form a new simplex and the response is measured for the combination of factor levels given by 4. Comparing the responses for the points 1, 2 and 4 will show that 1 now gives the worst response. The reflection process is repeated to give the simplex defined by 2, 4 and 5. The continuation of this process is shown in Fig. 7.10. It can be seen that no further progress is possible beyond the stage shown, since points 6 and 8 both give a worse response than 5 and 7.

In order to improve the performance of the simplex method various modifications have been (and still are being) proposed. The advance towards the maximum can be accelerated by using a simplex which can vary in size according to how the response for the new vertex in a simplex compares with the other vertices. Initially the simplex is large to give rapid progress towards the maximum; near the maximum it contracts to allow the maximum to be found accurately and to avoid the situation shown in Fig. 7.10.

The position of the new vertex of a simplex can be found by calculation rather than drawing — this is essential when there are more than two factors. The calculation is most easily set out as shown in Table 7.8. In this example there are 5

Table 7.8.   Simplex optimization

	A	B	Factors C	D	E	Response
Vertex 1	1.0	3.0	2.0	6.0	5.0	7
Vertex 2	6.0	4.3	9.5	6.9	6.0	8
Vertex 3	2.5	11.5	9.5	6.9	6.0	10
Vertex 4 (rejected)	2.5	4.3	3.5	6.9	6.0	6
Vertex 5	2.5	4.3	9.5	9.7	6.0	11
Vertex 6	2.5	4.3	9.5	6.9	9.6	9
(i) Sum (excluding vertex 4)	14.50	27.40	40.00	36.40	32.60	
(ii) Sum/n (excluding vertex 4)	2.90	5.48	8.00	7.28	6.52	
(iii) Rejected vertex (i.e. 4)	2.50	4.30	3.50	6.90	6.00	
(iv) Displacement = (ii) − (iii)	0.40	1.18	4.50	0.38	0.52	
(v) Vertex 7 = (ii) + (iv)	3.30	6.66	12.50	7.66	7.04	

factors and hence the simplex has 6 vertices. The response for vertex 4 is the lowest and so this vertex is to be replaced. The co-ordinates of the centroid of the vertices which are to be *retained* are found by (i) summing the co-ordinates for the retained vertices and (ii) dividing by the number of factors, n. The displacement of the new point from the centroid is given by (iv) = (ii) − (iii), and the co-ordinates of the new vertex, vertex 7, by (v) = (ii) + (iv). If the simplex is to be varied in size then the values in row (iv) are multiplied by a suitable scaling factor.

An obvious question in using the simplex method is the choice of the initial simplex. If this is taken as a *regular* figure in n dimensions, then the positions taken by the vertices in order to produce such a figure will depend on the scales used for the axes. As with the method of steepest ascent these scales should be chosen so that unit change in each factor gives roughly the same change in response. If there is insufficient information to achieve this, the difference between the highest and lowest feasible value of each factor can be represented by the same distance. The choice of the size of the initial simplex is discussed in a paper by Yarbro and Deming (see bibliography), who show that the size of the initial simplex is not so critical if it can be expanded or contracted as the method proceeds. Yarbro and Deming quote an algorithm which can be used to calculate the initial positions of the vertices; one vertex is normally positioned at the currently accepted levels of the factors.

It can be seen that in contrast to a factorial design the number of experiments required in the simplex method does not increase rapidly with the number of factors. For this reason all factors which may be thought to have a bearing on the response should be included in the optimization, since doing so will not greatly increase the number of experiments required to define the optimum.

Once an optimum has been found, the effect on the response when one factor is varied while the others are held at their optimum levels can be investigated for each factor in turn. This procedure can be used to check the optimization. It also indicates how important deviations from the optimum level are for each factor: the sharper the response peak in the region of the optimum the more critical any variation in factor level.

Simplex optimization has been used with success in many areas of analytical chemistry, e.g. atomic-absorption spectrometry, gas chromatography, colorimetric methods of analysis, plasma spectrometry, and use of centifugal analysers in clinical chemistry. When an instrument is interfaced with a microcomputer, the results of simplex optimization can be used to initiate automatic improvements in the instrument variables.

For response surfaces with more than one peak, such as that shown in Fig. 7.11,

Level of Y

Level of X

Fig. 7.11.   Contour diagram showing two localized maxima.

both types of optimization described may locate a local optimum such as A rather that the true optimum B. Starting in a different region and seeing whether the same optimum conditions are obtained provides some means of checking this point. Optimization methods provide information about only *some* parts of the response surface, since their object is to maximize the response with the least effort. If a complete picture of the response surface is required, more elaborate experiments covering the whole range of the variables are needed.

7.13 PATTERN RECOGNITION

The advance of automation has made possible the rapid collection of large amounts of data and the development of microcomputers has made sophisticated processing of these data possible. The name **chemometrics** has been coined to describe mathematical and statistical methods designed to extract useful chemical information from chemical data. Optimization is one field covered by the umbrella of chemometrics; another is the field of pattern recognition, to which we now turn.

One consequence of automation is that many variables can be determined simultaneously for the same sample, for example in clinical chemistry, chromatography and atomic-emission analysis. One use of such data in analytical chemistry is significance testing, e.g. determining whether an oil-spill comes from a particular source by using the intensities and/or frequencies of the many maxima obtained in the infrared or fluorescence spectra of such samples. Another use is classification, e.g. dividing the stationary phases used in gas–liquid chromatography into groups with similar properties by using the retention properties of a variety of solutes with different chemical properties. In both cases pattern recognition methods use the available data simultaneously rather than sequentially. The set of measurements which is used to characterize the sample is called a **pattern**. When only two variables are measured for each sample the pattern can be represented graphically by a point as shown in Fig. 7.12 where the co-ordinates of the point are the values taken by the

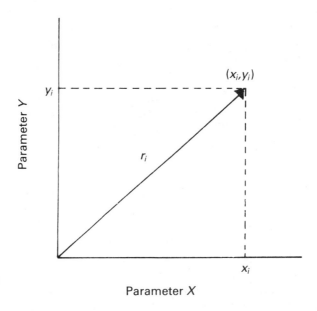

Fig. 7.12.   A pattern represented in pattern space; x_i and y_i are the values of the parameters X and Y; r_i is the pattern vector.

two variables. The point can also be defined by a vector, drawn to it from the origin and known as a **pattern** or **data vector**; the co-ordinate system is known as the **pattern space**. The basis of all pattern recognition methods is that pattern vectors for similar samples lie close together in the pattern space, forming **clusters**. In two dimensions this clustering can easily be detected by the human eye. However, when more than two variables are measured, graphical representation is no longer possible: if n variables are measured, each sample will be represented by a point in n-dimensional space and mathematical methods are needed to detect clustering. One such method, known as **principal component analysis** (PCA), allows the pattern vectors to be projected onto a plane in such a way that as little information as possible is lost. Figure 7.13 illustrates the principle of the method in the situation where there are just two variables. The diagram shows two well defined groups of pattern vectors and two possible lines onto which points could be projected, thus reducing the number of dimensions of the display from two to one. The projection onto line A would preserve as much distinction as possible between the two groups, whereas projection onto line B would obscure the difference. In this case a projection would be made onto line A. Lines A and B represent linear combinations of the original variables. Similarly, principal component analysis allows a projection to be made from n to 2 dimensions, in such a way as to preserve as much of the difference between the pattern vectors as possible. Again the axes of the plane onto which the projection is made represent linear combinations of the original variables. In many cases, only some of the original variables figure significantly in these linear combinations, and thus the variables which contribute most to the variation between the pattern vectors can be identified. The remaining variables, which convey little extra information, can be rejected. Indeed, if too many variables are used a random and chemically meaningless separation into classes may result, or the 'noise' from the superfluous variables may obscure the existence of classes.

Principal component analysis is closely related to another method which has become increasingly popular of late, viz. **principal components regression** (PCR). The latter is one of a family of methods which extends the principles of calibration and regression methods to situations where two or more components of a material can be determined simultaneously, by using two or more sets of instrument responses. Such methods have proved remarkably successful even when (as often happens in spectroscopy) the observed signals overlap strongly, or when the properties of one or more of the components of the mixture are not known in advance. In such cases, the role of PCR is again to reduce the number of components involved in the regression calculations. Other **multivariate regression** methods, including **partial least squares** (PLS) regression, have also proved effective, and are readily performed using personal computers and commercially available software. These and other methods require a knowledge of matrix algebra for detailed understanding, and are described in detail in the texts listed at the end of this chapter.

There are essentially two different approaches to pattern recognition, depending on whether it is being used for significance testing or classification. In the former case the classes into which a sample may fall are known and the purpose is to enable a sample of unknown class to be classified by its pattern. In the case of the oil-spill example, a sample may have come from one of several different sources: each source forms a class and the purpose of pattern recognition is to identify the class and hence the source of the sample. In the second case the purpose is to see whether the patterns obtained fall into natural groups; in this case there is no prior knowledge of the classes to be expected. These two different approaches to pattern recognition are sometimes called **supervised** and **unsupervised learning**, respectively. They use different methods to which we now turn.

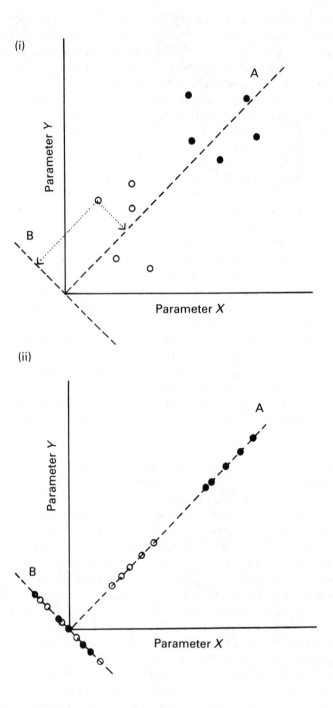

Fig. 7.13.   Projection of points from two dimensions to one dimension; (i) shows the original
points; (ii) shows their projections onto two different axes.

7.14 SUPERVISED LEARNING METHODS

The simplest situation is that in which a sample may belong to one of two classes: this is known as **binary classification**. It is illustrated in Fig. 7.14 for the two-variable case. In order to find a criterion or **classifier** for separating the two classes, a data set of patterns for samples with known class membership is used. This data set is divided (by using random numbers) into two parts called the **training set** and the **prediction set**. The training set is used to find the position of a plane, the **decision plane**, which separates the two classes. A simple way of doing this is to use the plane of symmetry between the centres of gravity of the two classes. This means that an unknown pattern vector is assigned to the class which has its centre of gravity closest to the vector. The distance, d, between two points in n-dimensional space with co-ordinates $(x_1, x_2, \ldots, x_n)$ and $(y_1, y_2, \ldots, y_n)$ is usually taken as the Euclidian distance, calculated by using the expression:

$$d^2 = \sum_{i=1}^{n} (x_i - y_i)^2$$

This method is satisfactory if the classes form compact clusters but, if they do not, more effective separation can be obtained by using more sophisticated methods. One of these is the **learning machine method**. This is an iterative procedure using negative feed-back. Starting with the plane of symmetry between the two centres of gravity as the decision plane, the patterns of the training set are tested one by one. If an incorrect classification is obtained the decision plane is adjusted to rectify this. The method is called the learning machine method because it 'learns' from mistakes in classification what adjustment is required in the position of the decision plane. Provided that two classes *can* be separated by a plane, this procedure eventually

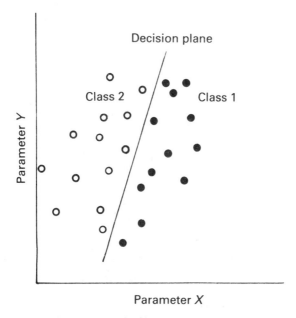

Fig. 7.14.   Two classes separated by a decision plane.

leads to the position of such a plane. If the classes are *not* separable by a plane, then the least-squares method is preferable. This method minimizes a sum of squares which estimates the errors in classification. Determining the position of the decision plane can also be treated as an optimization problem, solvable by the simplex method: the decision plane can be defined by a vector orthogonal to it and the components of this vector are the factors to be varied in level. The response which is maximized is the proportion of patterns from the training set which are correctly classified.

When the position of the decision plane has been fixed, the effectiveness of the decision plane in separating the two classes can be tested by using the remainder of the data set, i.e. the prediction set: each sample is classified by its pattern and the percentage correctly classified gives an objective estimate of the efficiency of the classifier.

In practice, before the pattern recognition procedure is applied, the variables (which may be measured in different numerical values) should be scaled so that the variables with the largest values do not dominate. A preliminary attempt may also be made to weight the variables so as to optimize the classification. After testing of the classifier these weighting factors may be further revised.

When there are more than two classes to be separated, a series of binary classifiers may be used. An alternative method which is conceptually very simple is the **K-nearest neighbour (KNN) method**. This method also has the advantage that it can be used when the classes cannot be separated by a plane, as illustrated in Fig. 7.15. In the simplest form of separation each sample is classified as in the same class as its nearest neighbour. Alternatively the K nearest neighbours (where K is a small integer) can be taken and class membership is decided by a voting scheme. An

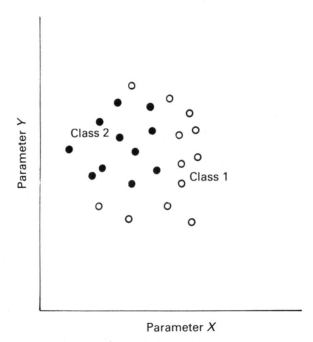

Fig. 7.15.   Two classes which cannot be separated by a plane.

odd value is normally taken for K to avoid ties. In a more sophisticated version different weightings can be given to the neighbours, depending on their relative distance. With this method no division of the data set into training and prediction sets is necessary — all the data can be used to estimate the efficiency of the method.

The approaches to pattern recognition so far described have taken into account only the distance between points and not their relative orientation in space. Such methods are non-parametric, i.e. no model is proposed for the distribution in pattern space of the samples within a class. In contrast the commercially available SIMCA computer program (due to Wold) uses the data set to form a mathematical model of each class. It is then possible to calculate for each class the probability that an unknown sample belongs to that class.

Some of the fields in which supervised pattern recognition methods have been applied are diagnostic tests in clinical chemistry, forensic science, the identification of sources of pollutants, food chemistry and the prediction of molecular structure from spectra.

7.15 UNSUPERVISED LEARNING METHODS

The reader will remember that these are methods which are used to decide whether a set of patterns divides naturally into groups. There are a large number of methods of searching for clusters in pattern space. One approach groups the pair of points which are closest together and replaces them by a new point half-way between them. This procedure is repeated and if continued indefinitely will group all the points together as shown by the steps in Fig. 7.16. In these diagrams the pairs of points which have

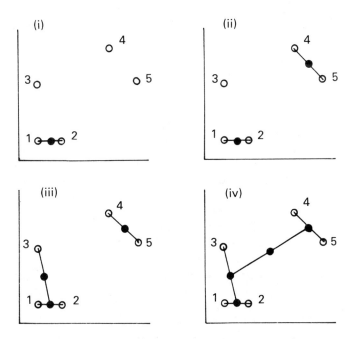

Fig. 7.16.   Stages in the clustering of patterns: ○ represents an original point, ● represents the single point replacing a pair of points.

been paired together and replaced by a single point are joined by a line. Thus at any stage, groups of points which are classed together are joined. The successive stages of the grouping can be shown on a **dendrogram** as in Fig. 7.17. The vertical axis shows

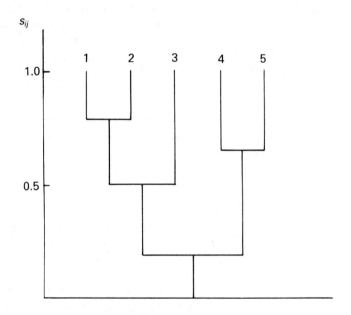

Fig. 7.17.   A dendrogram for the points in Fig. 7.16.

the **similarity**, s_{ij} between the two points i and j as calculated from:

$$s_{ij} = 1 - d_{ij}/d_{max}$$

where d_{ij} is the distance between the points i and j and d_{max} is the maximum distance between any two points. Obviously, the similarity takes values from 1 (for coincident points) to 0 (for the two points furthest apart). The stage at which the grouping is stopped, which determines the number of groups in the final classification, should be decided on a chemical basis. An opposite approach treats all the patterns as one group initially and then subdivides it.

As mentioned above, unsupervised learning methods have been used to classify the many phases used in gas–liquid chromatography. A small preferred set of phases with different characteristics can then be selected by taking one phase from each class. Another application is the classification of antibiotics in terms of their activity against various types of bacteria in order to elucidate the relationship between biological activity and molecular structure.

It will be seen that pattern recognition can be used to establish relationships which might otherwise be concealed in a mass of data. It is a method which has only fairly recently been applied in analytical chemistry: its potential as a means of data analysis remains to be fully evaluated. A number of computer packages are available besides SIMCA. One such is ARTHUR (developed by Kowalski) which contains programs for many methods, including those described here.

BIBLIOGRAPHY

G. E. P. Box, W. G. Hunter and J. S. Hunter, *Statistics for Experimentalists,* Wiley, New York, 1978. Gives a detailed treatment of complete and fractional factorial designs and the method of steepest ascent.
B. E. Cooper, *Statistics for Experimentalists,* Pergamon Press, Oxford, 1975. Gives further details of the mathematics of ANOVA.
D. R. Cox, *Planning of Experiments,* Wiley, New York, 1958. A non-mathematical treatment intended for the experimental worker. Gives further detail about Latin squares.
J. C. Davis, *Statistics and Data Analysis in Geology,* Wiley, New York, 1973. Gives a very clear account of principal component analysis.
B. R. Kowalski (ed.), *Chemometrics: Theory and Application,* American Chemical Society, Washington, 1977. Contains articles on SIMCA and ARTHUR.
L. Kryger, *Talanta,* 1981, **28**, 871. A survey of the application of pattern recognition methods to chemistry, including many references to original papers.
D. L. Massart, A. Dijkstra and L. Kaufman, *Evaluation and Optimization of Laboratory Methods and Analytical Procedures,* Elsevier, Amsterdam, 1978. A valuable source of information on all topics in this chapter. Includes many examples drawn from the literature and gives a large number of further references.
K. Varmuza, *Pattern Recognition in Chemistry,* Springer-Verlag, Berlin, 1980. Gives a detailed treatment of the methods and applications of pattern recognition.
L. A. Yarbro and S. N. Deming, *Anal. Chim. Acta,* 1974, **73**, 391. Discusses the choice of the initial simplex.
D. Betteridge, A. P. Wade and A. G. Howard, *Talanta,* 1985, **32**, 709, 723. Good source of references on simplex methods; compares various approaches.
D. L. Massart, B. G. M. Vandeginste, S. N. Deming, Y. Michotte, and L. Kaufman, *Chemometrics; a Textbook,* Elsevier, Amsterdam, 1988. An invaluable summary of modern methods, with numerous references to more detailed sources.
S. N. Deming and S. L. Morgan, *Experimental Design: a Chemometric Approach,* Elsevier, Amsterdam, 1987. Broader in its coverage than the title implies: contains a useful appendix on matrix algebra.
H. Martens and T. Naes, *Multivariate Calibration,* John Wiley, Chichester, 1989.

EXERCISES

(1) In an interlaboratory collaborative experiment on the determination of arsenic in coal, samples of coal from three different regions were sent to each of three laboratories. Each laboratory performed a duplicate analysis on each sample with the results shown below (measurements in μg/g).

		Laboratory	
Sample	1	2	3
A	5.1, 5.1	5.3, 5.4	5.3, 5.1
B	5.8, 5.4	5.4, 5.9	5.2, 5.5
C	6.5, 6.1	6.6, 6.7	6.5, 6.4

Verify that there is no significant laboratory–sample interaction, and test for significant differences between the laboratories.

(2) If the response at vertex 7 in the example on simplex optimization (p. 186) is found to be 12, which vertex should be rejected in forming the new simplex and what are the co-ordinates of the new vertex?

(3) Four standard solutions were prepared, each containing a calculated 16.00% (by weight) of chloride. Three titration methods, each with a different technique of end-point determination, were used to analyse each standard solution. The order of the experiments was randomized. The results for the chloride found (% w/w) are shown below:

Solution	A	Method B	C
1	16.03	16.13	16.09
2	16.05	16.13	16.15
3	16.02	15.94	16.12
4	16.12	15.97	16.10

Test whether there are significant differences between (a) the concentration of chloride in the different solutions and (b) the results obtained by the different methods.

(4) Samples of four different boiler feeds were analysed for nitrovin content by two different methods, each in duplicate. The results are shown in the table below (concentrations in mg/kg). Test whether the results obtained by the two different methods differ significantly.

Feed	HPLC method	Spectrophotometric method
1	9.3, 10.0	9.0, 10.2
2	9.3, 12.3	9.0, 11.8
3	12.8, 12.7	12.9, 12.0
4	11.9, 11.8	11.6, 12.1

(Adapted from M. J. Gliddon, C. Cordon, and G. M. Parnham, *Analyst*, 1983, **108**, 116)

(5) The effective temperature, calculated from the absorbance ratio of two spectral lines, for tin vapour was compared with the actual temperature under different experimental conditions in a heated quartz tube. The results are shown in the table below (values in °C). Calculate the main and interaction effects.

		Effective temperature		
		Purge gas		Purge gas Ar + 1% O_2
Purge time (sec)		20	60	20	60
Actual temperature (°C)	900	843	826	852	855
	1000	908	873	908	908

(Adapted from B. Welz and M. Melcher, *Analyst*, 1983, **108**, 213)

(6) Mercury is lost from solutions stored in polypropylene flasks by combination with adsorbed tin. The absorbance of a standard aqueous solution of mercury stored in such flasks was measured for two levels of the following factors:

Factor	Low	High
A — agitation of flask	Absent	Present
C — cleaning of flask	Once	Twice
T — standing time	1 hour	18 hours

The following results were obtained. Calculate the main and interaction effects.

Combination of factor levels	Absorbance
1	0.099
a	0.084
c	0.097
t	0.076
ac	0.082
ta	0.049
tc	0.080
atc	0.051

(Adapted from A. Kuldvere, *Analyst*, 1982, **107**, 179)

(7) The percentage recovery of lead in spiked samples in the presence of 'low' and 'high' concentrations of nitric acid and lanthanum was investigated by electrothermal atomic-absorption spectrometry; the means of duplicate analyses are given below. Calculate the main and interaction effects. If the residual mean square is 4 (with 4 degrees of freedom), test whether the interaction effect is significant.

	Lanthanum concentration (g/100 ml)	
Nitric acid concentration (% v/v)	0	0.10
0.1	64	90
1.0	77	96

(Adapted from M. P. Bertenshaw, D. Gelsthorpe and K. C. Wheatstone, *Analyst*, 1982, **107**, 163).

Solutions to exercises

CHAPTER 1

1. Laboratory A has achieved a mean result of 41.9 g/l., (very close to the correct value) and a small spread of results — all the values lie between 41.1 and 42.5 g/l. These results are thus accurate and precise: both random and systematic errors are small. Laboratory B has achieved the same mean value, but the spread of the results is much larger (range 39.8–43.9 g/l.), so although there are apparently no systematic errors (i.e. the mean result is accurate) there are large random errors (i.e. the data are very imprecise). Similar considerations show that laboratory C has produced precise but inaccurate results, (mean 43.2 g/l.), and that the results from laboratory D are both imprecise and inaccurate. Laboratory E has produced a set of results that look both precise and accurate, with the exception of the final value. In practice, this value might be tested as an outlier (see Chapter 3): if the tests showed that it could, with a reasonable level of confidence, be rejected, the remaining results would be quite similar to those of laboratory A.

2. The second set of six results obtained by Laboratory A has the same mean result as the first set, confirming that this laboratory produces results with no significant systematic error. In the second set the spread is greater (range 40.8–43.3 g/l.), illustrating the difference between within-day or within-batch precision (repeatability) and between-day or between-batch precision (reproducibility).

3. Monoclonal antibody populations are homogeneous, hence the number of binding sites per molecule must be a whole number — obviously two in this case. The results are thus precise, but show clear evidence of systematic error which produces low values.

4. (i) The lactate concentration in human blood varies quite widely from one healthy subject to another (ca. 5–20 mg/100 ml), and also varies to a lesser extent within one subject from time to time. If the degree of the latter type of variation is

studied, accuracy will not be very important but precise measurements will be necessary — the experimental errors must be negligible compared with the variations for the individual. If a single measurement is made, to check whether a subject does or does not fall within the 'normal range' for blood lactate, less precision will be required, but a large systematic error might lead to a wrong diagnosis.

(ii)   The uranium content of ores is studied with a view to economically viable extraction of the element. High precision will not therefore be necessary, but substantial systematic errors (positive or negative) might lead to economically disastrous decisions!

(iii)   In this analysis speed is clearly essential, and neither precision nor accuracy is of great importance initially. As the poisoned patient recovers, the plasma level of the drug may be repeatedly monitored to confirm that it is decreasing: since a trend is being studied, precision is more important than accuracy.

(iv)   Here again, the main aim is to detect changes in the analysis result; as these changes may be quite small, good precision will be necessary to detect any trend, but accuracy will be less essential.

5. (i)   In this experiment the most likely source of error is that the sample taken is not representative of the ore as a whole, and may thus give a wholly misleading value for the bulk Fe content (sampling is discussed in Chapter 4). A systematic error will also rise if the reduction of Fe(III) to Fe(II) is incomplete, or if there is a large indicator error. Systematic errors other than the sampling error can be checked with the aid of a standard ore sample — such samples are commercially available and are supplied with a certified Fe value. An additional problem, not necessarily solved by using the standard sample, is the possibility that other elements in low oxidation states will be titrated by the ceric sulphate, giving a falsely high result for iron. Random errors in volumetric analysis are discussed in the earlier sections of this chapter.

(ii)   Apart from the systematic errors discussed in (i), incomplete chelate formation and/or extraction will present the major problems in this case. Again, such errors will be detectable with the aid of an ore sample of known Fe content. The colorimetric analysis will probably use a series of Fe standards and a graphical procedure to calculate the result and the random errors: such methods are further discussed in Chapter 5.

(iii)   Random errors in gravimetric analysis should be very small and the systematic errors directly associated with the weighing process can be minimized by careful technique (see Section 1.4). The most likely sources of errors in this determination are chemical, e.g. the problem of co-precipitation of other ions, and are discussed at length in the established textbooks of classical analysis.

CHAPTER 2

1.   Mean $= 0.077 \ \mu g/g$. Standard deviation $= 0.007 \ \mu g/g$. RSD $= 9\%$.

2.   Mean $= 5.163$. Standard deviation $= 0.027$. Number of degrees of freedom $= 6$; (i) $5.16_3 \pm 0.025$; (ii) $5.163 \pm 0.03_8$.

3.   Number of degrees of freedom $= 9$; 95% confidence limits are $20.9_2 \pm 0.3_4$ g/l.

4.   Number of degrees of freedom $= 49$; 95% confidence limits are $10.1_2 \pm 0.1_8$ ng/ml. To reduce the range of the confidence interval a larger sample is required and it is safe to assume that t will be close to 1.96. The sample size is calculated from $0.1 = 1.96 \times 0.64/\sqrt{n}$, giving $n = 157$.

5.   Mean $= 10.45\%$. Standard deviation $= 0.105\%$. Number of degrees of freedom $= 5$. The 95% confidence limits are $10.4_5 \pm 0.1_1$. The 99% confidence limits are $10.4_5 \pm 0.1_7\%$. (i) No; (ii) no. This suggests that the method gives a systematic error.

6.   Mean $= 0.2943$. Standard deviation $= 0.00468$. Number of degrees of freedom $= 9$. The 99% confidence limits are $0.294_3 \pm 0.004_8$.

7.   Mean $= 10.19$. Standard deviation $= 0.19$. Number of degrees of freedom $= 4$. The 95% confidence limits are $10.1_9 \pm 0.2_3$, i.e. 9.96–10.42. Since the confidence interval includes 10.00 there is no evidence of systematic error.

8.   Standard deviation $= 0.14$ mg. Weight of reagent required is 0.5 (4.9) g giving RSD $= 0.028$ (0.0029)%. RSD for volume of solvent $= 0.02\%$. RSD of molarity of solution $= 0.034$ (0.020)%. Values for reagent with formula weight 392 are given in brackets.

9.   Solubility $= \sqrt{(1.3 \times 10^{-10})} = 1.1_4 \times 10^{-5}M$. RSD of solubility product $= 7.7\%$, giving RSD of solubility $= 3.85\%$ and standard deviation $= 0.04_4 \times 10^{-5}M$.

CHAPTER 3

1.   The points lie approximately on a straight line, indicating that the data are drawn from a normal distribution.

2. $Q = 0.13/0.23 = 0.565$. Critical value $= 0.570$. The result is not significant at $P = 0.05$; there is no evidence that 2.07 is an outlier.

3.   Tomato: mean $= 772.6$, standard deviation $= 13.6$ μg/g. Cucumber: mean $= 780.9$, standard deviation $= 10.4$ μg/g. (i) A two-tailed test is required. $F_{6,6} = 1.70$. Critical value $= 5.82$ ($P = 0.05$). The variances do not differ significantly. (ii) Since the variances do not differ significantly a pooled estimate of standard deviation is

calculated and is 12.1 $\mu g/g$; $t = 1.28$. For a two-tailed test, critical value $= 2.18$ ($P = 0.05$), so the means do not differ significantly.

4.

Source of variation	Sum of squares	Degrees of freedom	Mean square
Between-samples	6365.7116	3	2121.9
Within-samples	162.0284	20	8.1
	6527.74	23	

In order to test whether the percentage of water recovered differs significantly at different depths, the between-sample variation is compared with the within-sample variation: $F_{3,20} = 262$. Critical value $= 3.098$ ($P = 0.05$, one-tail test). The value is significant, indicating a difference between the sample means. The sample means are 34.1, 45.4, 72.2, 70.4. The least significant difference $= 3.43$, calculated taking $t_{20} = 2.09$ ($P = 0.05$, two-tail test). All pairs of depths, except the deepest pair, differ significantly from each other.

5.   Use the chi-squared test with each expected frequency $= 5$. $\chi^2 = 16.8$. There are 9 degrees of freedom and the critical value is 16.92 ($P = 0.05$). There is insufficient evidence that some digits are preferred.

6. $t = 1.54, 1.60, 1.18, 1.60$. There are 7 degres of freedom and the critical value is 2.36 ($P = 0.05$, two-tail test). None of the means differs significantly from the certified value.

7.   The means are 55.875 and 50.25 mg/kg and the standard deviations are 0.8345 and 8.828 mg/kg.
   (a) $F_{7,7} = 112$. Critical value $= 3.79$ ($P = 0.05$, one-tail test). Variability significantly greater for longer boiling time.
   (b)   A pooled estimate of variance is not appropriate as the variances differ significantly; $t = 1.8$. The number of degrees of freedom, calculated from Eq. (3.5) is 7. Critical value $= 2.36$ ($P = 0.05$, two-tail). The means do not differ significantly.

(8) (a)   Expected frequencies $= 15.25$ and 45.75, assuming that the number of breakages is in proportion to the number of workers in each group, whereas the actual frequencies are 24 and 37; 1 degree of freedom. $\chi^2 = 5.95$ (Yates's correction applied). Critical value $= 3.84$ ($P = 0.05$). The first worker differs significantly from other three.
   (b)   For the last three workers the expected frequencies $= 12.33, 12.33, 12.33$ (i.e. 37/3). $\chi^2 = 2.81$; 2 degrees of freedom. Critical value $= 5.99$ ($P = 0.05$). The last three workers do not differ significantly from each other.

9.   First we must test whether the variances differ significantly: $F_{2,2} = 1.22$ — not

significant ($P = 0.05$, two-tail). Thus a pooled estimate of variance can be calculated and is equal to 0.1685; $t = 0.54$. Critical value $= 2.78$ ($P = 0.05$, two-tail). Mean results do not differ significantly.

10.   In this example a paired t-test is appropriate as both methods are applied to each sample: testing the differences between the pairs removes the effect of any variation between the samples. Differences are 0.051, 0.030, -0.030 mg/ml. These differences have mean $= 0.017$ mg/ml, standard deviation $= 0.042$ mg/ml. There are 2 degrees of freedom and $t = 0.7$. Critical value $= 4.3$ ($P = 0.05$, two-tail). Results do not differ significantly.

11.   Men: mean $= 40.0$ g/l, standard deviation $= 2.78$ g/l. Women: mean $= 43.2_5$ g/l., standard deviation $= 3.06$ g/l. First we must test whether the variances differ significantly: $F_{7,7} = 1.21$. The critical value is 4.995 ($P = 0.05$, two-tail) so the variances do not differ significantly. The pooled estimate of s.d. is 2.92; $t = 2.2$. Critical value $= 2.14$ ($P = 0.05$, two-tail). The means differ significantly.

12.   A paired t-test is appropriate. The mean and s.d. of the differences are 1.18 and 0.84 respectively; $t = 3.4$. Critical value $= 2.57$ ($P = 0.05$, two-tail). The results differ significantly.

CHAPTER 4

1.   Scheme 1: $\sigma^2 = (4/2) + (10/5) = 4$
      Scheme 2: $\sigma^2 = 4/(2 \times 3) + 10/3 = 4$

If $S = $ cost of sampling and $A = $ cost of analysis then:

$$\text{cost of scheme 1 : cost of scheme 2} = (5S + 2A) : (3S + 6A)$$

This ratio is greater than 1 if $S : A$ exceeds $2 : 1$.

2.   The ANOVA calculations are summarized below:

Source of variation	Sum of squares	d.f.	Mean square
Between-days	332.92	3	111
Within-days	26.00	8	3.25
Total	358.92	11	

$F_{3,8} = 34$. Critical value $= 4.066$ ($P = 0.05$) showing that the mean concentrations differ significantly. Estimate of sampling variance $= (111 - 3.25)/3 = 35.9$.

3.   The ANOVA calculations are summarized below:

Source of variation	Sum of squares	d.f.	Mean square
Between-sample	0.0024916667	3	0.0008306
Within-sample	0.0014	8	0.000175
Total	0.0038916667	11	

$F_{3,8} = 4.746$. The critical value $(P = 0.05)$ is 4.066, showing that the between-sample mean square is too great to be explained by measurement variation.

The within-sample mean square, 0.000175, gives an estimate of the measurement variance, σ_0^2.

Estimate of sampling variance, $\sigma_1^2 = (0.0008306 - 0.000175)/3 = 0.000219$.

Variance of the mean for scheme 1 $= 0.000175/4 + 0.000219/6 = 0.00008025$.

Variance of the mean for scheme 2 $= [0.000175/(2 \times 3)] + 0.000219/3 = 0.0001022$.

4. From each sample an estimate of σ^2 is made, giving: 2.607, 0.847, 3.017, 3.633, 6.417, 1.927. The average of these values is 3.077. Thus the estimate of $\sigma = \sqrt{3.077} = 1.753$.

The warning lines are at $50 \pm 2 \times 1.753/\sqrt{4}$, i.e. at 48.25 and 51.75.

The action lines are at $50 \pm 3 \times 1.753/\sqrt{4}$, i.e. at 47.37 and 52.63.

5. Simple calculation shows that the mean Cd concentrations in samples A and B, as determined by the 15 laboratories, are 7.01 and 7.75 ppm respectively. When the point (7.75, 7.01) is plotted on the two-sample chart, and vertical and horizontal lines are drawn through it, the four quadrants generated contain (reading clockwise from the $+$, $+$ quadrant) 6, 1, 7 and 1 points. This result is typical of a collaborative trial of this type, and suggests that the large differences between the 15 laboratories are mainly caused by systematic errors. It is noteworthy that only four laboratories (7, 9, 12, and 13) generate results that are fairly close to the means for both samples, whereas laboratories 3 and 10 have large positive and negative biases respectively for both samples. Perpendicular distances from the experimental points to the 45° line drawn through (7.75, 7.01) are measures of the random errors of the laboratories concerned, and it is clear that laboratories 4 and 14 may have large random errors in their results.

To calculate the components of the variation, the usual table is set up:

Laboratory	A	B	D	T
1	8.8	10.0	− 1.2	18.8
2	3.8	4.7	− 0.9	8.5
3	10.1	12.1	− 2.0	22.2
4	8.0	11.0	− 3.0	19.0
5	5.0	4.7	0.3	9.7
6	5.2	6.4	− 1.2	11.6
7	6.7	8.7	− 2.0	15.4
8	9.3	9.6	− 0.3	18.9
9	6.9	7.5	− 0.6	14.4
10	3.2	2.8	0.4	6.0
11	9.7	10.4	− 0.7	20.1
12	7.2	8.3	− 1.1	15.5
13	6.5	6.8	− 0.3	13.3
14	9.7	7.2	2.5	16.9
15	5.0	6.0	− 1.0	11.0

From the D and T columns in the table, it is easy to calculate that the estimate of σ_0^2 = 0.793, and that the estimate of σ^2 = 11.027. The experimental $F_{14,14}$ value is thus 11.027/0.793 = 13.905: this is much larger than the tabulated value (ca. 2.48, obtained from Table A.2 by interpolation), showing that the systematic errors are indeed significant. They can be estimated from Eq. (4.3), the result being σ_1^2 = 5.117.

CHAPTER 5

1. Application of Eq. (5.2) to the data gives a product-moment correlation coefficient, r, of -0.8569. This value suggests a considerable negative correlation between the mercury level and the distance from the polarograph. Its significance can be tested by using Eq. (5.3), which yields a t-value of 3.33. Comparison of this value with the tabulated value of t at the 95% confidence level and $n-2 (= 4)$ degrees of freedom (t = 2.13) shows that the relationship is indeed significant at this confidence level. Two cautions, however: (1) this method tests only linear correlation, whereas in the laboratory we might expect that the mercury level would decrease with increasing distance from the polarograph in an inverse square manner; (2) correlation is *not* the same as causation. The calculation shows that there is a relationship between distance from the polarograph and mercury concentration — it does *not* show that the polarograph is the source of the mercury contamination. The contamination might arise (for example) from a sink or another piece of equipment close to the polarograph.

2. In this case Eq. (5.2) shows that $r = 0.99982$. This is apparently strong evidence for linear correlation, *but* inspection of the data shows clearly that a curve is more suitable in this case. Each y (absorbance) value is greater than its predecessor, but by a slowly decreasing increment as x increases. The curvature can also be demonstrated by studies of the residuals (cf. exercise 9). In this case, the errors caused by treating the data as a linear calibration plot would be very small.

3. Application of Eqs. (5.4) and (5.5) gives $b = 0.0252$, and $a = 0.0021$ respectively. Other important results include: $\bar{x} = 15$; $\bar{y} = 0.380$; $\Sigma x_i^2 = 2275$; and $\sum_i (x_i - \bar{x})^2 = 700$. The individual y residuals $(y_i - \hat{y}_i)$ are found to be $+0.0009$, -0.0009, -0.0028, $+0.0104$, -0.0074, -0.0062, $+0.0060$. These residuals sum (as expected) to zero, and the sum of their squares is 0.000247. Application of Eq. (5.6) shows that $s_{y/x} = 0.00703$. Then Eq. (5.7) gives $s_b = 0.000266$, and the use of t (5 degrees of freedom; 95% confidence level) $= 2.57$ yields for b confidence limits of 0.0252 ± 0.0007. Similarly, Eq. (5.8) yields $s_a = 0.00479$, with confidence limits for a of 0.0021 ± 0.0123.

4. (a) It is easy to show, using the a and b values calculated in exercise 3, that a y

value of 0.456 corresponds to an x value of 18.04 ng/ml, and that $(y_0 - \bar{y})^2 = 0.00578$. The application of Eq. (5.9) then gives $s_{x_0} = 0.300$, and the corresponding confidence limits are given by $18.04 \pm (2.57 \times 0.300) = 18.04 \pm 0.77$ ng/ml.

(b) In this case it is clearly necessary to see whether the absorbance value 0.347 is an outlier. From Eq. (3.8) the calculated value of Q is $(0.347 - 0.314)/(0.347 - 0.308) = 0.846$. At the 95% confidence level, this just exceeds the tabulated Q value of 0.829, so the value 0.347 *can* be rejected. This leaves three results, with a mean absorbance of 0.311, corresponding to a concentration of 12.28 ng/ml. In this case the value of s_{x_0} is given by Eq. (5.10) $[m = 3; (y_0 - \bar{y})^2 = 0.00476]$ as 0.195, the confidence limits thus being $12.28 \pm (2.57 \times 0.195) = 12.28 \pm 0.50$ ng/ml.

5.   Armed with the results of exercise 3 above, this calculation is very simple. The limit of detection is defined as the concentration yielding a signal exceeding the background absorbance signal by three standard deviations, the background being given by a (0.0021) and the standard deviation by $s_{y/x}$ (0.00703). The absorbance signal at the limit of detection is thus $0.00211 + (3 \times 0.00703) = 0.0232$, and the limit of detection, calculated from the slope and intercept of the regression line, is 0.84 ng/ml.

6.   The usual linear regression calculation gives $a = 0.2569$ and $b = 0.005349$. The ratio of these two numbers gives the concentration of the test sample $-$ 48.0 ng/ml. Also, $\bar{x} = 35$, $\bar{y} = 0.4441$, $\Sigma(x_i - \bar{x})^2 = 4200$, and $\Sigma(y_i - \hat{y}_i)^2 = 0.00001364$. Thus $s_{y/x} = 0.003693$, and s_{xE}, given by Eq. (5.12) is 0.9179. The confidence limits for the concentration are thus $48.03 \pm (2.45 \times 0.9179) = 48.0 \pm 2.2$ ng/ml.

7.   The data provided allow a weighted regression calculation, the mean y values and standard deviations being:

x:	0	10	20	30	40	50
y:	4.0	21.2	44.6	61.8	78.0	105.2
s:	0.7_1	0.8_4	0.8_9	1.6_4	2.2_4	3.0_3

The first step is to calculate the unweighted regression line, the slope and intercept being 1.982 and 2.924 respectively. The concentrations corresponding to fluorescence intensities of 15 and 90 units are thus 6.09 and 43.9 ng/ml respectively. In the usual way (cf. exercise 3 above), $s_{y/x}$ is calculated to be 2.991. Since $\Sigma(x_i - \bar{x})^2$ is readily calculated as 1750 and $\bar{y}$ as 52.47, the s_{x_0} values for the two concentrations can be shown [Eq. (5.9)] to be 1.767 in each case. By use of a t value of 2.78, the 95% confidence limits for these values are thus given by 6.1 ± 4.9 and 43.9 ± 4.9 ng/ml respectively.

To calculate the weighted regression line, it is first necessary to calculate the weight for each point by using Eq. (5.13). In order of increasing x_i value, these weights are found to be 2.227, 1.591, 1.418, 0.417, 0.224, and 0.122 — note that the sum of these numbers is 5.999, i.e. 6 if we allow for rounding errors. This is an important check to make. By use of these weights and Eqs. (5.14) and (5.15) the slope and intercept of the weighted regression line are found to be 1.964 and 3.483 respectively. From this line, the concentrations corresponding to fluorescence intensitites of 15 and 90 units are 5.87 and 44.1 ng/ml respectively. Note that these values, and those for the slope and intercept, are all similar to those calculated for the unweighted line, despite the considerable scatter of the experimental points. The crucial difference between the use of the weighted and unweighted calculation methods only becomes apparent when Eq. (5.16) is used to estimate the standard deviations of the estimated concentrations. To use this equation, it is first necessary to estimate w_0 values for the two y_0 values under study. Inspection of the data suggests that weights of 1.8 and 0.18 respectively should be appropriate for the y_0 values of 15 and 90. Use of these weights gives $s_{x_{0w}}$ values of 0.906 and 2.716 respectively. Multiplication by the t value 2.78 gives the confidence limits for the two concentrations as 5.9 ± 2.5 and 44.0 ± 7.6 ng/ml respectively.

8. Here, the 'new' method is clearly the ISE one, plotted therefore on the y-axis: the gravimetric technique is the reference or standard procedure and is thus plotted on the x-axis. The regression line calculated in this way has a slope of 0.963, and an intercept of 4.48, the correlation coefficient being 0.970. This last value is clearly highly significant [$t = 11.29$ — Eq. (5.3)] but the confidence limits of the slope and intercept must also be calculated. The usual calculations [Eqs. (5.6)–(5.8)] in conjunction with a tabulated t value of 2.31 (8 degrees of freedom; 95% confidence) show that the confidence limits for the slope and the intercept are respectively 0.96 ± 0.200 and 4.5 ± 20.1. Since these 95% confidence intervals clearly include 1 and 0 respectively, we must conclude that this regression line indicates good agreement between the two analytical methods. Nonetheless, there is a residual feeling of unease about the data. Calculation shows that 7 of the y-residuals are negative (including 6 in a single sequence) and only 3 positive. There is also a suggestion that the agreement between the two methods is very close for most of the samples, but very poor for a few of them; this could be caused by speciation problems. A good analyst might conclude that more data are needed (more are actually supplied in the paper cited).

9. If these data are plotted graphically (remember that this should *always* be done), there is some indication that the calibration graph is linear up to $0.7 - 0.8$ absorbance units, and curved thereafter. Unweighted linear regression calculations using all six points yield a correlation coefficient of 0.9936, and y-residuals (in order of increasing x) of -0.07, -0.02, $+0.02$, $+0.06$, $+0.07$, and -0.07 (rounded to 2 places of decimals). The trend in these values is obvious and confirms that we are dealing with a curve. The sum of squares of the residuals is 0.0191. Confirmation that the last point should be omitted from the linear range is obtained by repeating the regression calculation with only the five points. The correlation coefficient then increases to 0.9972 and the y-residuals are -0.04, 0, $+0.02$, $+0.04$, and -0.02 respectively (sum of squares 0.0040). The y-residual for the sixth point with respect to this second straight line is very large by comparison: -0.22.

This process can then be repeated for the fifth point, i.e. the straight line fitted to only the first four points is calculated. Again there is an improvement in the correlation coefficient (to 0.9980), a reduction in the values of the y-residuals (-0.02, $+0.01$, $+0.01$, and -0.01), and a large residual for the fifth point under test (-0.14). It can thus be concluded that the fifth point can also be excluded from the linear range of the method. Similar tests applied to the fourth point indicate by contrast that it probably should be part of the linear range, and an analyst could use linear regression methods in the absorbance range 0–0.6 with confidence.

CHAPTER 6

1.   The mean is 9.96 ml, and the median (the average of the second and third ranked measurements) is 9.90 ml. The Q-test shows that the value 10.20 cannot quite be rejected at $P = 0.05$. If it *were* rejected the mean would move to 9.88 ml and the median to 9.89 ml; the median is insensitive to the presence of outliers.

2. (a)   Use of the sign test: compared with the median, the experimental values give signs of $- + 0 + - + + + +$. In effect we thus have 8 readings, 6 with $+$ signs. The chance of six signs out of 8 being the same is 0.29, much greater than 0.05. So we can retain the null hypothesis that the data could have come from a symmetrical population with median sulphur content of 0.10%.
   (b)   In the signed rank test, the zero value is again neglected. The remaining data compared with the median give ranked differences of $-0.01, 0.01, 0.01 - 0.02$, $0.02, 0.02, 0.04, 0.07$. The signed ranks, taking into account the ties, are thus $-2, 2$, $2, -5, 5, 5, 7, 8$. The negative ranks total $(-)7$ but, at $P = 0.05$, the critical value for $n = 8$ is only 3. Again, therefore, the null hypothesis is retained.
   (c)   To use a range test, we calculate the mean ($\bar{x}$), which is 0.1156, and the range (w), 0.09. Then T_I is given by $|\bar{x} - V|/w = 0.173$. The critical value in this case is 0.255, so for the third time the null hypothesis is retained. Contrary to superficial appearances, perhaps, the sample of fuels cannot be said to fail the specification.

3.   This problem can also be solved by using the sign test and the signed rank test. If the e.i.d. values are subtracted from r.i.d. values, the signs of the differences are $+ - + + + + + + 0 +$. In effect we have 9 results, with 8 $+$ signs and one $-$ sign. The probability of 8 signs out of nine being the same is only 0.04, so at the significance level $P = 0.05$ we can reject the null hypothesis: the results *are* significantly different. In the signed rank test, the magnitude of the negative rank is only 2.5. This is well below the critical value of 5, so again the null hypothesis is rejected, and the significant difference between the two sets of results is confirmed. Note, however, that the results of the signed rank test should be interpreted with caution if there are a lot of tied ranks, as there are both in this question and the previous one.

4.   If the values are arranged in ascending order, the median is found to be 23.5. The individual values, in order of use of the carbon rods, thus have signs + + + − − − − − + +. There are thus three runs. But for $N = M = 5$ the critical value is 3, so the given sequence might well have been a random one.

5.   This problem is easily solved by the U-test. If the 'beer' values are expected to be greater than the 'lager' values, the number of lager values that are greater than the individual beer values should be small. It is actually 4.5, counting 0.5 for the one tie that occurs. The critical value in a *one-tailed* test is 5, so we can (just) reject the null hypothesis at the $P = 0.05$ level and say that one particular lager does produce lower blood alcohol levels than one particular beer! However, the result in Tukey's quick test, $T = 5.5$, just fails to reach the critical value of 6 (one-tailed, $P = 0.05$). So the results must be treated with caution — more data are needed. The Siegel–Tukey test for spread gives a test statistic of 10, far above the critical value of 2. So the null hypothesis of similar spreads must be retained.

6.   The ranks are as follows:

Instrument:	A	B	C	D	E	F	G
Student ranking:	3	1	5	4	7	6	2
Staff ranking:	5	3	6	2	4	7	1
d:	− 2	− 2	− 1	2	3	− 1	1
d^2:	4	4	1	4	9	1	1

Thus ρ is $1 - (6 \times 24)/(7 \times 48) = 0.571$. This is well below the critical value of 0.786 ($P = 0.05$ and $n = 7$), so the students and staff do not seem to agree — the correlation is not significant.

7.   If the x values are the distances, and the y values the mercury levels, then Theil's method gives $a = 2.575$, and $b = -0.125$. These values are astonishingly close to the (parametric) 'least squares' values of 2.573 and -0.122.

8.   To test $\bar{x} = 1.0$ and $\sigma = 0.2$, we write $z = (x - 1.0)/0.2$, and obtain 1.5, 2.5, -1.5, -0.5, 0, 0.5, -1.0, 4.0, -3.0, and 1.5. Plotted against the cumulative distribution function for the standard normal distribution, these values gives a trace that is fairly symmetrical, but much broader than the standard curve, indicating that the assumed mean is not very far wrong, but the estimated standard deviation is far

too low. The maximum difference is about $+0.335$ at $z = 1.5$. This is higher than the critical value of 0.262, confirming that the null hypothesis must be rejected, and that the values do not fit this particular normal distribution. To test for another normal distribution, we calculate 1.08 and 0.41 as our estimates of the mean and standard deviation respectively. These give z values of 0.54,. 1.02, -0.93, -0.44, -0.20, 0.05, -0.68, 1.76, -1.66, and 0.54. The maximum difference is this case is only ca. 0.11 at $z = 0.54$, so the null hypothesis is retained; the data fit this normal distribution very well.

9.   This problem is solved by replacing the actual nickel concentrations with rank values, with the following results:

Sample			Rank values			
1	1	8	5	15	4	6
2	2	17	10.5	13	3	7
3	14	18	9	10.5	12	16

Note that one tie occurs, and is dealt with in the normal way. The sums of the three sets of ranks are: $R_1 = 39$, $R_2 = 52.5$, $R_3 = 79.5$. These three numbers add up to 171, equal to $\frac{1}{2}N(N+1)$, where $N = 18$, the total number of measurements. The value of chi-squared for this example is thus given by:

$$\chi^2 = \frac{12}{342 + 18}\left(\frac{1521}{6} + \frac{2756.25}{6} + \frac{6320.25}{6}\right) - 3\,(18 + 1)$$

$$= 4.97$$

This is less than the critical value for χ^2 of 5.99 ($P = 0.05$, 2 degrees of freedom — Table A.5). We must therefore retain the null hypothesis, and conclude that there is no significant difference between the nickel levels in the three oils.

CHAPTER 7

1.   An example of two-way ANOVA with replication.

Source of variation	Sum of squares	d.f.	Mean square
Between-row	5.06778	2	2.53
Between-column	0.18778	2	0.0939
Interaction	0.10222	4	0.0256
Residual	0.36500	9	0.0406
Total	5.72278	17	

The residual mean square is greater than the interaction mean square so there is no significant laboratory–sample interaction. To test whether there is a significant difference between laboratories the between-column mean square is compared with the residual mean square, giving:

$$F_{2,9} = 0.0939/0.0406 = 2.31$$

The critical value of $F_{2,9}(P = 0.05)$ is 4.256, so the calculated value is not significant.

2.   Vertex 1 should be rejected in forming the new simplex.

	Factors					Response
	A	B	C	D	E	
Vertex 2	6.0	4.3	9.5	6.9	6.0	8
Vertex 3	2.5	11.5	9.5	6.9	6.0	10
Vertex 5	2.5	4.3	9.5	9.7	6.0	11
Vertex 6	2.5	4.3	9.5	6.9	9.6	9
Vertex 7	3.3	6.7	12.5	7.7	7.0	12
Sum	16.8	31.1	50.5	38.1	34.6	
Sum ÷ n	3.4	6.2	10.1	7.6	6.9	
Rejected vertex (i.e.1)	1.0	3.0	2.0	6.0	5.0	
Displacement	2.4	3.2	8.1	1.6	1.9	
Vertex 8	5.8	9.4	18.1	9.2	8.8	

(All values in the table above are given to 1 decimal place).

3.   An example of two-way ANOVA without replication. First subtract 16 from each value.

Source of variation	Sum of squares	d.f.	Mean square
Between-row	0.011092	3	0.00370
Between-column	0.012014	2	0.00601
Residual	0.028186	6	0.00470
Total	0.051292	11	

   (a)   The between-solution (i.e. row) mean square is smaller than the residual mean square and so is not significant.
   (b)   Comparing the between-method (i.e. column) and residual mean squares gives $F_{2,6} = 1.28$, which is not significant ($P = 0.05$).

4.   An example of two-way ANOVA with replication.

Source of variation	Sum of squares	d.f.	Mean square
Between-row	20.8768749	3	6.959
Between-column	0.14062595	1	0.141
Interaction	0.11187458	3	0.037
Residual	9.92499971	8	1.241
Total	31.0543752	15	

There is no significant interaction or difference between methods since in both cases the corresponding mean square is less than the residual mean square.

5. Using the following notation:

G = purge gas (Ar — low, Ar + O_2 — high)
T = purge time (20 sec — low, 60 sec — high)
Q = silica cell temperature (900°C — low, 1000°C — high)

the calculated effects are:

Single-factor	Effect
G	18.25
T	− 12.25
Q	55.25

Two-factor	
GT	13.75
TQ	− 5.25
GQ	− 0.75

Three-factor	
GTQ	3.75

6.

Single-factor	Effect
A	− 0.0215
C	0.0005
T	− 0.0265

Two-factor	
AC	− 0.0005
CT	0.0025
AT	− 0.0065

Three-factor	
ACT	− 0.0005

7.   Using the following notation:

   N = nitric acid concentration ($0.1M$ — low, $1.0M$ — high)
   L = lanthanum concentration ($0M$ — low, $0.1M$ — high)

the calculated effects are:

Single-factor	Effect	Sum of squares	d.f.
N	9.5	180.5	1
L	22.5	1012.5	1
Interaction			
NL	3.5	24.5	1

To test for significant interaction the interaction mean square is compared with the residual mean square, giving $F_{1,4} = 24.5/4 = 6.125$. The critical value of $F_{1,4}$ is 7.709 ($P = 0.05$), so there is no significant interaction.

Appendix 1   Summary of statistical tests

The following tables summarize the statistical tests described in this book. The tests are grouped according to their application, with brief details and comments in each case, and a reference to the page on which further details may be found. It is hoped that this will help the reader in deciding on the statistical test appropriate to a particular problem.

(1) Tests on averages

Test name or test statistic	Used to test whether	Refer to page	Comments
t	(a) a sample mean differs from a standard value (b) the means of two samples differ (c) paired samples differ	53 55 58	Assumes that populations are normal
Sign test	(a) a sample median differs from a standard value (b) paired samples differ	146 147	Non-parametric
Wilcoxon signed rank test	(a) a sample median differs from a standard value (b) paired samples differ	149 150	Non-parametric but assumes that the population is symmetrical; requires $n \geqslant 5$ for one-tailed and $n \geqslant 6$ for two-tailed tests
Wilcoxon rank sum test	(a) the medians of two samples differ	152	Non-parametric; minimum sample sizes: one-tailed test, $n_1 = n_2 = 3$; two-tailed test, $n_1 = 3$, $n_2 = 5$
Mann–Whitney U-test	(a) the medians of two samples differ	153	Simpler version of Wilcoxon rank sum test
Tukey's quick test	(a) the medians of two samples differ	153	Very simple alternative to the rank sum and U-tests — needs no statistical tables
Kruskal–Wallis test	(a) the medians of three or more samples differ	155	Non-parametric; uses χ^2 method in modified form
Friedman test	(a) matched samples differ	156	Non-parametric; χ^2 method in modified form

(2) Tests on spread

Test name or test statistic	Used to test whether:	Refer to page	Comments
F	variances of two samples differ	60	Assumes normality
Siegel–Tukey test	spreads of two samples differ	154	Non-parametric; modified from a Wilcoxon rank sum test
(3) Tests on runs			
Wald–Wolfowitz	a sequence is random	148	For minimum numbers of data points see Table A.8
(4) Tests on correlation			
Product-moment correlation coefficient, r	data are linearly correlated	104	Assumes that errors are normally distributed
Spearman rank correlation coefficient, ρ	data are linearly correlated	158	Simpler to calculate than r; non-parametric
(5) Goodness of fit tests			
χ^2 (chi-squared)	observed frequencies differ from expected frequencies	71	Requires total sample ≥ 50 and individual frequencies ≥ 5
Kolmogorov–Smirnov	data conform to a specified distribution	165	Can be used for small samples; particularly useful in testing for normality

Appendix 2   Statistical Tables

The following tables are presented for the convenience of the reader, and for use with the simple statistical tests, examples and exercises in this book. They are presented in a format that is compatible with the needs of analytical chemists: the significance level $P = 0.05$ has been used in most cases, and it has been assumed that the number of measurements available is fairly small. Except where stated otherwise, these abbreviated tables have been taken, with permission, from *Elementary Statistics Tables* by Henry R. Neave, published by George Allen & Unwin Ltd. (Tables A.1–A.3, A.5–A.7, and A.12–A.16). The reader requiring statistical data corresponding to significance levels and/or numbers of measurements not covered in the tables is referred to these sources.

Table A.1.   The t-distribution

Value of t for a confidence interval of	90%	95%	98%	99%
Critical value of $|t|$ for P values of	0.10	0.05	0.02	0.01
Number of degrees of freedom				
1	6.31	12.71	31.82	63.66
2	2.92	4.30	6.96	9.92
3	2.35	3.18	4.54	5.84
4	2.13	2.78	3.75	4.60
5	2.02	2.57	3.36	4.03
6	1.94	2.45	3.14	3.71
7	1.89	2.36	3.00	3.50
8	1.86	2.31	2.90	3.36
9	1.83	2.26	2.82	3.25
10	1.81	2.23	2.76	3.17
12	1.78	2.18	2.68	3.05
14	1.76	2.14	2.62	2.98
16	1.75	2.12	2.58	2.92
18	1.73	2.10	2.55	2.88
20	1.72	2.09	2.53	2.85
30	1.70	2.04	2.46	2.75
50	1.68	2.01	2.40	2.68
∞	1.64	1.96	2.33	2.58

The critical values of $|t|$ are appropriate for a *two*-tailed test. For a *one*-tailed test the value is taken from the column for *twice* the desired P-value, e.g. for a one-tailed test, $P = 0.05$, 5 degrees of freedom, the critical value is read from the $P = 0.10$ column and is equal to 2.02.

Table A.2. Critical values of F for a one-tailed test $(P = 0.05)$

ν_1	1	2	3	4	5	6	7	8	9	10	12	15	20
ν_2													
1	161.4	199.5	215.7	224.6	230.2	234.0	236.8	238.9	240.5	241.9	243.9	245.9	248.0
2	18.51	19.00	19.16	19.25	19.30	19.33	19.35	19.37	19.38	19.40	19.41	19.43	19.45
3	10.13	9.552	9.277	9.117	9.013	8.941	8.887	8.845	8.812	8.786	8.745	8.703	8.660
4	7.709	6.944	6.591	6.388	6.256	6.163	6.094	6.041	5.999	5.964	5.912	5.858	5.803
5	6.608	5.786	5.409	5.192	5.050	4.950	4.876	4.818	4.772	4.735	4.678	4.619	4.558
6	5.987	5.143	4.757	4.534	4.387	4.284	4.207	4.147	4.099	4.060	4.000	3.938	3.874
7	5.591	4.737	4.347	4.120	3.972	3.866	3.787	3.726	3.677	3.637	3.575	3.511	3.445
8	5.318	4.459	4.066	3.838	3.687	3.581	3.500	3.438	3.388	3.347	3.284	3.218	3.150
9	5.117	4.256	3.863	3.633	3.482	3.374	3.293	3.230	3.179	3.137	3.073	3.006	2.936
10	4.965	4.103	3.708	3.478	3.326	3.217	3.135	3.072	3.020	2.978	2.913	2.845	2.774
11	4.844	3.982	3.587	3.357	3.204	3.095	3.012	2.948	2.896	2.854	2.788	2.719	2.646
12	4.747	3.885	3.490	3.259	3.106	2.996	2.913	2.849	2.796	2.753	2.687	2.617	2.544
13	4.667	3.806	3.411	3.179	3.025	2.915	2.832	2.767	2.714	2.671	2.604	2.533	2.459
14	4.600	3.739	3.344	3.112	2.958	2.848	2.764	2.699	2.646	2.602	2.534	2.463	2.388
15	4.543	3.682	3.287	3.056	2.901	2.790	2.707	2.641	2.588	2.544	2.475	2.403	2.328
16	4.494	3.634	3.239	3.007	2.852	2.741	2.657	2.591	2.538	2.494	2.425	2.352	2.276
17	4.451	3.592	3.197	2.965	2.810	2.699	3.614	2.548	2.494	2.450	2.381	2.308	2.230
18	4.414	3.555	3.160	2.928	2.773	2.661	2.577	2.510	2.456	2.412	2.342	2.269	2.191
19	4.381	3.522	3.127	2.895	2.740	2.628	2.544	2.477	2.423	2.378	2.308	2.234	2.155
20	4.351	3.493	3.098	2.866	2.711	2.599	2.514	2.447	2.393	2.348	2.278	2.203	2.124

ν_1 = number of degrees of freedom of the numerator and ν_2 = number of degrees of
freedom of the denominator.

Table A.3. Critical values of F for a two-tailed test $(P = 0.05)$

ν_1	1	2	3	4	5	6	7	8	9	10	12	15	20
ν_2													
1	647.8	799.5	864.2	899.6	921.8	937.1	948.2	956.7	963.3	968.6	976.7	984.9	993.1
2	38.51	39.00	39.17	39.25	39.30	39.33	39.36	39.37	39.39	39.40	39.41	39.43	39.45
3	17.44	16.04	15.44	15.10	14.88	14.73	14.62	14.54	14.47	14.42	14.34	14.25	14.17
4	12.22	10.65	9.979	9.605	9.364	9.197	9.074	8.980	8.905	8.844	8.751	8.657	8.560
5	10.01	8.434	7.764	7.388	7.146	6.978	6.853	6.757	6.681	6.619	6.525	6.428	6.329
6	8.813	7.260	6.599	6.227	5.988	5.820	5.695	5.600	5.523	5.461	5.366	5.269	5.168
7	8.073	6.542	5.890	5.523	5.285	5.119	4.995	4.899	4.823	4.761	4.666	4.568	4.467
8	7.571	6.059	5.416	5.053	4.817	4.652	4.529	4.433	4.357	4.295	4.200	4.101	3.999
9	7.209	5.715	5.078	4.718	4.484	4.320	4.197	4.102	4.026	3.964	3.868	3.769	3.667
10	6.937	5.456	4.826	4.468	4.236	4.072	3.950	3.855	3.779	3.717	3.621	3.522	3.419
11	6.724	5.256	4.630	4.275	4.044	3.881	3.759	3.664	3.588	3.526	3.430	3.330	3.226
12	6.554	5.096	4.474	4.121	3.891	3.728	3.607	3.512	3.436	3.374	3.277	3.177	3.073
13	6.414	4.965	4.347	3.996	3.767	3.604	3.483	3.388	3.312	3.250	3.153	3.053	2.948
14	6.298	4.857	4.242	3.892	3.663	3.501	3.380	3.285	2.209	3.147	3.050	2.949	2.844
15	6.200	4.765	4.153	3.804	3.576	3.415	3.293	3.199	3.123	3.060	2.963	2.862	2.756
16	6.115	4.687	4.077	3.729	3.502	3.341	3.219	3.125	3.049	2.986	2.889	2.788	2.681
17	6.042	4.619	4.011	3.665	3.438	3.277	3.156	3.061	2.985	2.922	2.825	2.723	2.616
18	5.978	4.560	3.954	3.608	3.382	3.221	3.100	3.005	2.929	2.866	2.769	2.667	2.559
19	5.922	4.508	3.903	3.559	3.333	3.172	3.051	2.956	2.880	2.817	2.720	2.617	2.509
20	5.871	4.461	3.859	3.515	3.289	3.128	3.007	2.913	2.837	2.774	2.676	2.573	2.464

ν_1 = number of degrees of freedom of the numerator and ν_2 = number of degrees of
freedom of the denominator.

Appendices

Table A.4.   Critical values of Q ($P = 0.05$) for a two-sided test

Sample size	Critical value
4	0.831
5	0.717
6	0.621
7	0.570
8	0.524
9	0.492
10	0.464

Taken from E. P. King, *J. Am. Statist. Assoc.*, 1958, **48**, 531, by permission of the American Statistical Association.

Table A.5.   Critical values of χ^2 ($P = 0.05$)

Number of degrees of freedom	Critical value
1	3.84
2	5.99
3	7.81
4	9.49
5	11.07
6	12.59
7	14.07
8	15.51
9	16.92
10	18.31

Table A.6.   Random numbers

02484	88139	31788	35873	63259	99886	20644	41853	41915	02944
83680	56131	12238	68291	95093	07362	74354	13071	77901	63058
37336	63266	18632	79781	09184	83909	77232	57571	25413	82680
04060	46030	23751	61880	40119	88098	75956	85250	05015	99184
62040	01812	46847	79352	42478	71784	65864	84904	48901	17115
96417	63336	88491	73259	21086	51932	32304	45021	61697	73953
42293	29755	24119	62125	33717	20284	55606	33308	51007	68272
31378	35714	00941	53042	99174	30596	67769	59343	53193	19203
27098	38959	49721	69341	40475	55998	87510	55523	15549	32402
66527	73898	66912	76300	52782	29356	35332	52387	29194	21591
61621	52967	40644	91293	80576	67485	88715	45293	59454	76218
18798	99633	32948	49802	40261	35555	76229	00486	64236	74782
36864	66460	87303	13788	04806	31140	75253	79692	47618	20024
10346	28822	51891	04097	98009	58042	67833	23539	37668	16324
20582	49576	91822	63807	99450	18240	70002	75386	26035	21459
12023	82328	54810	64766	58954	76201	78456	98467	34166	84186
48255	20815	51322	04936	33413	43128	21643	90674	98858	26060
92956	09401	58892	59686	10899	89780	57080	82799	70178	40399
87300	04729	57966	95672	49036	24993	69827	67637	09472	63356
69101	21192	00256	81645	48500	73237	95420	98974	36036	21781
22084	03117	96937	86176	80102	48211	61149	71246	19993	79708
28000	44301	40028	88132	07083	50818	09104	92449	27860	90196
41662	20930	32856	91566	64917	18709	79884	44742	18010	11599
91398	16841	51399	82654	00857	21068	94121	39197	27752	67308
46560	00597	84561	42334	06695	26306	16832	63140	13762	15598

Table A.7. The sign test

The table uses the binomial distribution with $p = 0.5$ to give the probabilities of r or less successes for $n = 4$–15. These values correspond to a one-tailed sign test and should be doubled for a two-tailed test.

n	$r = 0$	1	2	3	4	5	6	7
4	0.063	0.313	0.688					
5	0.031	0.188	0.500					
6	0.016	0.109	0.344	0.656				
7	0.008	0.063	0.227	0.500				
8	0.004	0.035	0.144	0.363	0.637			
9	0.002	0.020	0.090	0.254	0.500			
10	0.001	0.011	0.055	0.172	0.377	0.623		
11	0.001	0.006	0.033	0.113	0.274	0.500		
12	0.000	0.003	0.019	0.073	0.194	0.387	0.613	
13	0.000	0.002	0.011	0.046	0.133	0.290	0.500	
14	0.000	0.001	0.006	0.029	0.090	0.212	0.395	0.605
15	0.000	0.000	0.004	0.018	0.059	0.151	0.304	0.500

Table A.8. The Wald–Wolfowitz runs test

N	M	At $P = 0.05$, the number of runs is significant if it is:	
		Less than	Greater than
2	12–20	3	NA
3	6–14	3	NA
3	15–20	4	NA
4	5–6	3	8
4	7	3	NA
4	8–15	4	NA
4	16–20	5	NA
5	5	3	9
5	6	4	9
5	7–8	4	10
5	9–10	4	NA
5	11–17	5	NA
6	6	4	10
6	7–8	4	11
6	9–12	5	12
6	13–18	6	NA
7	7	4	12
7	8	5	12
7	9	5	13
7	10–12	6	13
8	8	5	13
8	9	6	13
8	10–11	6	14
8	12–15	7	15

Adapted from F. S. Swed and C. Eisenhart, *Ann. Math. Statist.*, 1943, **14**, 66, by permission of the Institute of Mathematical Statistics. The test cannot be applied to data with N, M smaller than the given numbers, or to cases marked NA.

Table A.9.   Wilcoxon signed rank test

Critical values for the test statistic at $P = 0.05$.

n	One-tailed test	Two-tailed test
5	0	NA
6	2	0
7	3	2
8	5	3
9	8	5
10	10	8
11	13	10
12	17	13
13	21	17
14	25	21
15	30	25

The null hypothesis can be rejected when the test statistic is $\leqslant$ the tabulated value. NA indicates that the test cannot be applied.

Table A.10.   Wilcoxon rank sum test; Mann–Whitney U-test

Critical values for U or the lower of T_1 and T_2 at $P = 0.05$

n_1	n_2	One-tailed test	Two-tailed test
3	3	0	NA
3	4	0	NA
3	5	1	0
3	6	2	1
4	4	1	0
4	5	2	1
4	6	3	2
4	7	4	3
5	5	4	2
5	6	5	3
5	7	6	5
6	6	7	5
6	7	8	6
7	7	11	8

The null hypothesis can be rejected when U or the lower T value is $\leqslant$ the tabulated value. NA indicates that the test cannot be applied.

Table A.11.   The Spearman rank correlation coefficient

Critical values for ρ at $P = 0.05$

n	One-tailed test	Two-tailed test
5	0.900	1.000
6	0.829	0.886
7	0.714	0.786
8	0.643	0.738
9	0.600	0.700
10	0.564	0.649
11	0.536	0.618
12	0.504	0.587
13	0.483	0.560
14	0.464	0.538
15	0.446	0.521
16	0.429	0.503
17	0.414	0.488
18	0.401	0.472
19	0.391	0.460
20	0.380	0.447

Table A.12.   The Kolmogorov test for normality

Critical values for one-tailed and two-tailed tests at $P = 0.05$. The appropriate value is compared with the maximum difference between the experimental and theoretical cumulative frequency curves, as described in the text.

n	One-tailed test	Two-tailed test
3	0.367	0.376
4	0.345	0.375
5	0.319	0.343
6	0.297	0.323
7	0.280	0.304
8	0.265	0.288
9	0.252	0.274
10	0.241	0.262
11	0.231	0.251
12	0.222	0.242
13	0.215	0.234
14	0.208	0.226
15	0.201	0.219
16	0.195	0.213
17	0.190	0.207
18	0.185	0.202
19	0.181	0.197
20	0.176	0.192

Index